Y0-DBU-576

A New Era For Irrigation

Committee on the Future of Irrigation
in the Face of Competing Demands

Water Science and Technology Board

Commission on Geosciences, Environment, and Resources

National Research Council

NATIONAL ACADEMY PRESS
Washington, D.C. 1996

NOTICE: The project that is the subject of this report was approved by the Governing Board of the National Research Council, whose members are drawn from the councils of the National Academy of Sciences, the National Academy of Engineering, and the Institute of Medicine. The members of the board responsible for the report were chosen for their special competences and with regard for appropriate balance.

This report has been reviewed by a group other than the authors according to procedures approved by a Report Review Committee consisting of members of the National Academy of Sciences, the National Academy of Engineering, and the Institute of Medicine.

Support for this project was provided by the The Ford Foundation Grant No. 930-0484, The Irrigation Association, National Water Research Institute, the Bureau of Reclamation Grant No. 3-FG-81-19130, and U.S. Department of Agriculture, Agricultural Research Service Grant No. 59-0700-2-147.

Library of Congress Cataloging-in-Publication Data

A new era for irrigation/Committee on the Future of Irrigation in
the Face of Competing Demands, Water Science and Technology Board,
Commission on Geosciences, Environment, and Resources.
p. cm.
Includes bibliographical references and index.
ISBN 0-309-05331-5
1. Irrigation—United States. 2. Irrigation farming—United
States. I. National Research Council (U.S). Committee on the
Future of Irrigation in the Face of Competing Demands.
S616.U6N47 1996
631.5'87—dc20 96-25369

Original cover art by Sally Groom, Arlington, Virginia

Printed in the United States of America

COMMITTEE ON THE FUTURE OF IRRIGATION IN THE FACE OF COMPETING DEMANDS

WILFORD GARDNER, *Chair*, University of California, Berkeley
KENNETH FREDERICK, *Vice Chair*, Resources for the Future, Washington, D.C.
HEDIA ADELSMAN, State of Washington, Olympia, Washington
JOHN S. BOYER, University of Delaware, Lewes
CHELSEA CONGDON, Environmental Defense Fund, Boulder, Colorado
DALE F. HEERMANN, U.S. Department of Agriculture, Ft. Collins, Colorado
EDWARD KANEMASU, University of Georgia, Athens, Georgia
RONALD D. LACEWELL, Texas A&M University, College Station
LAWRENCE MacDONNELL, Sustainability Initiatives, Boulder, Colorado
THOMAS K. MacVICAR, MacVicar and Associates, Inc., Lake Worth, Florida
STUART T. PYLE, Consulting Engineer, Bakersfield, California
LESTER SNOW, San Diego County Water Authority, San Diego, California (through February 16, 1995)
CATHERINE VANDEMOER, U.S. Department of the Interior, Washington, D.C.
JAMES WATSON, Toro Company, Littleton, Colorado
JAMES L. WESCOAT, JR., University of Colorado, Boulder
HOWARD A. WUERTZ, Sundance Farms, Collidge, Arizona

Liaison from the Water Science and Technology Board

CAROLYN H. OLSEN, Brown and Caldwell, Pleasant Hill, California

National Research Council Staff

CHRIS ELFRING, Study Director, Water Science and Technology Board
ANITA A. HALL, Senior Project Assistant, Water Science and Technology Board

WATER SCIENCE AND TECHNOLOGY BOARD

COMMISSION ON GEOSCIENCES, ENVIRONMENT, AND RESOURCES

The National Academy of Sciences is a private, nonprofit, self-perpetuating society of distinguished scholars engaged in scientific and engineering research, dedicated to the furtherance of science and technology and to their use for the general welfare. Upon the authority of the charter granted to it by the Congress in 1863, the Academy has a mandate that requires it to advise the federal government on scientific and technical matters. Dr. Bruce Alberts is president of the National Academy of Sciences.

The National Academy of Engineering was established in 1964, under the charter of the National Academy of Sciences, as a parallel organization of outstanding engineers. It is autonomous in its administration and in the selection of its members, sharing with the National Academy of Sciences the responsibility for advising the federal government. The National Academy of Engineering also sponsors engineering programs aimed at meeting national needs, encourages education and research, and recognizes the superior achievements of engineers. Dr. William A. Wulf is interim president of the National Academy of Engineering.

The Institute of Medicine was established in 1970 by the National Academy of Sciences to secure the services of eminent members of appropriate professions in the examination of policy matters pertaining to the health of the public. The Institute acts under the responsibility given to the National Academy of Sciences by its congressional charter to be an adviser to the federal government and, upon its own initiative, to identify issues of medical care,research, and education. Dr. Kenneth I. Shine is president of the Institute of Medicine.

The National Research Council was organized by the National Academy of Sciences in 1916 to associate the broad community of science and technology with the Academy's purposes of furthering knowledge and advising the federal government. Functioning in accordance with general policies determined by the Academy, the Council has become the principal operating agency of both the National Academy of Sciences and the National Academy of Engineering in providing services to the government, the public, and the scientific and engineering communities. The Council is administered jointly by both Academies and the Institute of Medicine. Dr. Bruce Alberts and Dr. William A. Wulf are chairman and interim vice chairman, respectively, of the National Research Council.

Preface

The Committee on the Future of Irrigation in the Face of Competing Demands was asked to explore how irrigation might best make the transition into an era of increasing water scarcity. The charge resembles a description of the role of what a scientist does that appeared in the *House of Solomon*, written four centuries ago by Francis Bacon:

> And we do also declare natural divinations (forecasting by natural observation) of diseases, plagues, swarms of hurtful creatures, scarcity, tempests, earthquakes, great inundations, comets, temperature of the years, and diverse other things; and we give counsel thereupon, what the people shall do for the prevention and remedy of them. (*The New Atlantis*, 1597, p. 302)

In the committee's case, "divinations" about the future of irrigation range from thoughts on environmental sustainability and technological innovation to ponderings on global economic competitiveness. Like those long-ago scientists, we, too, are called on to offer counsel on how society might address these and other unforeseen challenges that irrigated agriculture may face or may pose for society.

The study was challenging because of this somewhat amorphous underpinning. We debated what future meant—5 years, 25 years, 100 years? We debated the meaning of the term "sustainable." We were awed at the diversity of agriculture in this nation and how difficult it is to generalize about issues. We wondered how we could add anything new and of substance to the many fine voices already addressing these problems. We struggled with how to provide conclusions, and even more difficult, recommendations, to the study's sponsors and the nation, from what is essentially a philosophical exploration.

In the end, I am proud to say that the committee's members have created an excellent document that explores this nation's relationship to irrigation in a thoughtful and thought-provoking way. I want to express my thanks to each of them for contributing so much time and energy to this project. I also would like to thank the study's sponsors—USDA's Agricultural Research Service, DOI's Bureau of Reclamation, the Irrigation Association, the National Water Research Institute, and the Ford Foundation—for their financial support and their willingness to seek outside guidance on sensitive issues. I would also like to thank the staff of the Water Science and Technology Board—Anita Hall for her diligent support work and Chris Elfring for her behind-the-scenes leadership.

No one can predict the future, but neither can we afford to ignore it. During periods of uncertainty, concerned citizens often look to "experts" to provide interpretation, guidance, and counsel. This committee was asked to peer into an uncertain future and provide advice about how best to prepare for whatever changes that future may bring. We of course can make no claims of omniscience or infallibility. Through listening, questioning, reading, and arguing with a broad cross section of people—especially farmers and others with hands-on experience in irrigation and related fields—the committee hoped to identify and analyze the range of key factors influencing how irrigation will evolve. We set out to explore the forces of change that affect the irrigation sector and to see how different regions are responding. From there, we hoped to find clues about where irrigation could and should head in the future. We hope that this report presents an accurate portrayal of what we learned and that despite our criticisms and calls for change we were able to convey our admiration for all that the practitioners of irrigation have contributed to society. We hope the next generation of farmers will respond to today's challenges with as much energy and innovation.

Wilford Gardner, *Chair*
Committee on the Future of Irrigation

Contents

Summary

Irrigation has played a vital role in the history of the United States, a role that extends far beyond the production of food and fiber. Irrigation was a driving force in the settlement of the American West. Its success spawned a culture and sparked an evolution of technology and myriad supporting institutions. These institutions have continued to support irrigation at great benefit to the nation, although not without costs to other water users, the environment, and social programs. The United States is now facing a time of changing public values and new demands, however, and irrigators feel a combination of pressures today unlike any time in the past. The availability of water has been, and is likely to remain, the principal determinant of the status of irrigation in the western United States and is becoming increasingly important to irrigation in eastern states as well. But the cost of water and demands on the resource are changing rapidly.

The special place of irrigation in American society is a result of a long and complex history that involves federal policies, individual entrepreneurial spirit, creativity, natural disasters, economics, and trading patterns. The remarkable development of irrigation was fueled by an enormous level of federal involvement, including engineering and financial assistance, but this role has diminished greatly in recent years. Growth no longer needs encouragement—the West is the most rapidly urbanizing part of the nation. And the role of agriculture in the nation's economy has changed. In 1900, 4 in 10 workers were engaged in farming; today, it is closer to 3 in 100. At the same time, some of the nation's priorities have shifted, including an increased concern for environmental issues, the need to compete in an increasingly global economy, and pressures from population growth and urbanization. Parks, nurseries, recreation areas, landscaping, lawns, and golf courses have become bona fide users of large quantities of water. Instream uses of water such as fishing, boating, fulfillment of treaty obligations with American Indians, and preservation of fish and wildlife habitat also have greater standing.

The underlying premise of this study is that given increasing competition for water supplies, changes in how water is managed, allocated, and valued are inevitable. Irrigation must be able to adapt to these conditions. Some of the pressures that affect the nature of change include:

- Water costs are rising, as is demand for water, and both trends are likely to continue.
- As the largest and most economically marginal user of water in the many water-scarce areas, irrigated agriculture is particularly vulnerable to changing water availability.
- The viability of farming on millions of irrigated acres is threatened by problems such as salinization of soils and dependence on nonrenewable water supplies.
- The quality of irrigation drainage or return flows often is sufficiently impaired as to limit the future reuse of that water for other purposes, including environmental uses.
- Irrigation systems and management will continue to evolve, moving toward advanced technologies that provide better water control.
- The ability of states, Indian tribes, and individual water users to market water will be central to increasing the flexibility of water allocation, whether for irrigation or nonirrigation uses, and thus is key to the future of irrigation in the United States.

THE CULTURE OF IRRIGATION

At one level, irrigation is simply the application of water to grow plants. At another level, it is the basis for an economy and a way of life. Irrigation made possible the highly intensive settlement of the western United States that otherwise would not readily support large populations. It has transformed the landscape, literally and figuratively. Thus there is a somewhat intangible, subjective dimension of irrigation that must be understood as we find ways to adapt and prepare for the future—the context in which change must occur, or what this report has called the "culture of irrigation."

Culture, as used here, refers to the "ideas, customs, skills, arts, etc. of a given people in a given period." Irrigation is a distinctive activity, one with its own history, own governmental policies, institutions, practices, and, historically, its own communities. The federal government facilitated this culture based on the central idea that a society and an economy could be built on irrigated agriculture as its base. It was a bold idea, ideally suited for the era of western expansion and exploitation, and it met with widespread support. The irrigation culture viewed itself as serving a larger national interest as well as providing a means of subsistence in an arid environment.

In today's increasingly urbanized society, evidence of a culture of irrigation is much less apparent, and other critical issues have emerged as priorities. But this traditional irrigation culture remains vibrant in many rural areas, and it continues to support economies. It is, however, a culture somewhat in retreat and on the defensive. Instead of a national symbol of progress and growth, irrigation is criticized for the pollution it produces and the subsidies that sustain it. The nation once supported the subsidization of agriculture (particularly irrigated agriculture) as a way of promoting the national interest—by using it to encourage settlement of the West and by stabilizing farmers' incomes and crop prices. It was also a way to provide a low cost supply of food and fiber to U.S. consumers. This is changing, however, and to compete effectively in the future, irrigators must change in ways that help overcome negative perceptions.

Cultural concerns influence irrigation systems and policies across the nation, but they have been neglected in scientific research and policy analysis. It would be an error to assess irrigation problems today without studying the full record of the experiences that created them and that might lead beyond them. Historical and cultural studies shed light on the knowledge systems of the present; they remind us that modern irrigation systems reflect complex social, economic, institutional, and technological influences. Understanding irrigation in its cultural context can help identify new approaches to problem-solving, combining new technologies and business practices with traditional technologies and approaches, as necessary, to respond to changing local, national, and global situations.

FORCES OF CHANGE AND RESPONSES

The principal determinants of the profitability of irrigated agriculture and, therefore, its future are: the overall state of the agricultural economy and markets for agricultural products; the benefits of irrigated farming relative to dryland farming (e.g., consistent high quality production); the cost and availability of water; pricing policy and the regulatory structure; available technology and management skills; the cost of other agricultural inputs such as labor, capital, and energy; environmental concerns and regulations; and the institutions that influence how water is used.

One method for anticipating the future of irrigation in the face of competing demands is to identify the forces of change that are affecting irrigators today and examine reactions to these forces. Key forces of change are competition over water supplies, changing economic conditions, changing values and policy objectives, and increasing environmental concerns. These forces can be addressed by actions at various levels, ranging from individual farms to the state, federal, and tribal institutions. Responses can take many forms, including developments in science, technology, and management and institutional and policy reforms. There is no clear delineation among these issues. In fact, there is extensive overlap and feedback among the forces of change and the responses to that change.

In terms of science and technology, responses have shifted away from construction of large-scale public works—dams and water delivery systems—toward improved on-farm irrigation systems that tend to reduce the total quantity of water that must be diverted from a stream for delivery to the farm. There is a trend toward adoption of microirrigation systems that apply water at a slow, carefully calibrated rate just below the soil surface. Researchers are also working to develop plant varieties that are better adapted to water stress. Although such efforts may help reduce the need for irrigation in the future, dramatic water savings from genetic engineering do not appear imminent.

From an institutional perspective, responses are occurring at a variety of levels. Federal policies are in a period of transition. The focus of federal policies affecting water use has shifted sharply over the past 25 years, from development of dams and other facilities to better use and management of existing facilities, diminishing subsidies, and increased environmental protection. At the same time, shrinking federal budgets make the future role of environmental programs and conservation subsidies (e.g., the conservation reserve program, which provides payments to farmers for leaving highly erodible land unplanted) increasingly uncertain. States, which set the rules governing allocation of water resources within their boundaries, are beginning to adopt changes in water laws and related review processes to encourage and facilitate voluntary transfers of water and water rights. The Bureau of Reclamation also has made efforts to accommodate voluntary transfers of Reclamation-supplied water. Even so, western water law—with its emphasis on "use it or lose it"—remains in need of revision to provide incentives for more efficient water use. Where economic incentives are lacking, voluntary conservation efforts may not be sufficient. Consequently, some states and local water districts are turning to regulatory approaches. In some instances, state law also is changing to reflect increased interest in protecting instream uses of water. In many areas, local, state, tribal, and federal institutions have turned to watershed approaches to address changing water demands. These watershed approaches offer important opportunities for irrigation interests to negotiate and resolve issues in a more integrated way.

EXAMPLES OF CHANGE AND RESPONSES

Although it is possible to describe the nature of irrigation and the issues with which irrigators and the industry must contend in general terms, it is more difficult to speak about the future without looking at irrigation as actually practiced in different regions. For example, while competition for water supplies and policies to protect environmental resources are issues affecting irrigation nationwide, the specifics of water supply problems and environmental restrictions are quite different in the Pacific Northwest than in the Texas High Plains.

Selected case studies are presented in Chapter 5 to illustrate these variations in problems and responses in four regions: the Great Plains, California, the Pacific Northwest, and Florida. Each presents different physical patterns, cultural patterns, functional economic relations, and jurisdictional relations. Physical differences are manifest in climate, hydrology, topography, and soils—which in turn influence certain irrigation practices, technology choices, public policy, and investments. Cultural differences affect choices of technologies and practices, the structure and philosophy of local and regional institutions, and responses to environmental regulation and changing public policy. Functional relations are the interconnections that shape the economic geography of the region, including factors such as local, regional, and global markets; labor supplies; availability of financial capital; and types of crops grown and related subsidy programs. Similarly, jurisdictional relations—the political and administrative entities with impact in the region—also affect how irrigation develops, what constraints apply, the context for solving environmental problems, and access to information, technical assistance, and technology.

FUTURE DIRECTIONS

No one can say with any degree of certainty how irrigated agriculture will change in the near or far term. We can, however, assert with considerable confidence that it will change. Few companies produce the same product in the same way they did 50 years ago, and agriculture is no exception. Irrigation attained its present stature because it was part of the nation's vision of how best to meet the needs of its citizenry. Whether one agrees or disagrees with the judgments and values that guided past decisionmakers, it is impossible not to admire the dedication with which that vision was put into action. The hope is that the current generation of decisionmakers and citizens can be as clear in their goals and as effective in designing a course to achieve them.

This committee has examined many factors that may influence the future of irrigation. These factors—especially competition for water; concerns over environmental impacts, including the potential impacts of climate change; the expansion of urban land uses, the globalization of the U.S. economy; the shifting roles of federal and state governments; and tribal economic development—will effect irrigation differently in different regions. Overall, the availability and cost of water are likely to remain the principal determinants of the extent of irrigation in the western United States; they are becoming increasingly important influences in the southern and eastern United States as well. From discussions with a wide range of people involved in irrigation and water use, field visits, study, and debate, the committee concludes[1]:

- Irrigation will continue to play an important role in the United States over the next 25 years, although certainly there will be changes in its character, methods, and scope. It is likely that irrigated acreage will decline overall, but the value of irrigated production will remain about the same because of shifts to higher-value crops.

- Given changing societal values and increasing competition for water, the amount of water dedicated to agricultural irrigation will decline. The availability and cost of water to the farmer are likely to remain the principal determinants of the extent of irrigation in the western United States; these factors are becoming increasingly important influences in the southern and eastern states as well.
- The economic forces driving irrigated agriculture increasingly will be determined by our ability to compete in global markets. This shift toward globalization, combined with reductions in protection and support for individual farmers, means that farmers will have to deal with increased levels of risk and uncertainty.
- The structure of irrigated agriculture will continue to shift in favor of large, well-financed, integrated, and diversified farm operations. Smaller, under-financed operations and those with less skilled managers will tend to decline.
- Many important federal, state, and local policies and institutions affecting irrigation were established in a different era, and they no longer meet contemporary societal needs. Changes in these policies and institutions are occurring to reflect changing economies, emerging values, and shifting policy priorities. Thus, for example, the Bureau of Reclamation is moving from a project construction agency to a water management agency. Innovation and flexibility will be needed, especially as direct federal support continues to diminish.
- In the past, the term irrigation effectively meant irrigation for agriculture. But the nature of irrigation has changed dramatically in the past two decades and will continue to change. Turf irrigation is now an important part of the irrigation industry, and irrigation for urban landscaping and golf courses in particular will continue to expand as urban populations increase.
- Advances in irrigation technology are necessary if both agricultural and turf irrigation are going to adapt to changing demands and changing supplies. The irrigation industry will need to play a larger role in technology development and dissemination as the federal government trims its support for these activities.
- Some portion of the water now in agricultural use will over time be shifted to satisfy environmental goals. In addition, there will be continued pressure to reduce environmental problems associated with irrigation, both agricultural and turf.
- Irrigation emerged as an individual and collective effort at the watershed level, and in many important respects its future will be determined in the watershed. The growth of locally driven watershed activities reflects a promising trend in water management.

Irrigation, to use a hydrologic metaphor, is at a watershed divide—a time in history where change is imminent. Irrigated agriculture must evolve to compete in a new era. It must adopt more efficient technologies and management strategies, develop more flexible institutional arrangements, and work cooperatively with other water users to allocate limited water resources equitably. Recommendations outlining various actions that might be undertaken at the federal, state, tribal, or local levels are discussed in detail in Chapter 6. In general, the recommendations address improved institutional arrangements, research and development of irrigation technologies and techniques, environmental protection, and the role of education and extension in disseminating innovation widely throughout the irrigation community. For example, the report discusses the need for the education and extension system to evolve to help farmers gain the skills needed to compete in an increasingly globalized economy. It notes that states will need to establish improved systems to facilitate the voluntary transfer of water among users. It also recommends that environmental regulation be flexible enough to deal with specific problems and locations, and involve incentive-based problems, investment credits, and similar tools that can enhance local- and regional-level environmental problem-solving.

There will continue to be an important role for the federal government in the future of irrigation, but this role, and the measures used to implement it, is changing. Federal support for research and development of new irrigation technologies will remain important, but with the continuing pressure to reduce federal expenditures, more leadership and funding for research and development will have to come from the private sector and through partnerships between irrigators, the private sector, and state and federal researchers. The federal government has important trust responsibilities in resolving the water rights of Indian tribes, and federal funding and commitment to this process is necessary to resolve unsettled tribal claims and reduce tensions over the future of tribal irrigation and the availability of tribal water, whether through transfers or other arrangements, for use by irrigators or other users.

Irrigation has served this nation well and will continue to provide benefits—food, fiber, and the support of rural communities. To continue in a new era, however, irrigation must evolve. Although the committee does not foresee explosive changes on the horizon, certainly the future will bring surprises, some perhaps dramatic. It is critical that the same resourcefulness that has made irrigation such an important economic and cultural activity over the past 100 years be brought to bear in the future.

NOTE

1. More detailed discussions of these conclusions, including implications for the future direction of irrigation, appear in Chapter 6.

1

The Future of Irrigation

Irrigated agriculture has played a critical role in the economic and social development of the United States. In terms of crop production, irrigated farms contribute proportionally more than nonirrigated farms: irrigated lands make up only about 15 percent of all harvested cropland yet they produce nearly 38 percent of the total crop value from the nation's agricultural lands (Bajwa et al., 1992). From a social perspective, irrigation served as the engine that drove western settlement, and today it supports local rural economies in the arid West and in wetter regions as well.

But the future of irrigation, and particularly the future of irrigated agriculture, in the United States is probably less clear today than at any other time during the past 50 years. The reasons for the uncertainties facing the practitioners of irrigation are neither surprising nor mysterious. Intense competition for water among an increasingly wide range of users, changing economics, increased environmental concerns, changing public values, and other trends in modern times are putting new pressures on irrigation.

No individual or group, regardless of wisdom and experience, can predict with assurance how, where, and when irrigation will be practiced in years to come, but it is imperative to at least consider these issues. In the best of worlds, it is better to anticipate problems than to react to crises and better to be proactive than simply to follow the path of least resistance. This is the rationale behind all efforts to anticipate the future, including this report.

Irrigation is an old art, probably one of the earliest agricultural practices. It arose out of necessity first in arid areas of the world, where irrigation was essential to ensure plant survival in the absence of timely precipitation. Archeological sites around the world provide ample evidence of centuries of irrigation and of highly successful economic and social systems built on the productivity that it enabled. History also bears record of irrigation-related problems and failures, and whether these are due to technological or political causes we can only surmise.

While it is not for this document to revisit the history of irrigated agriculture in the United States, some historical perspective is necessary to understand the present-day context of irrigation. The special place of irrigation today is a result of a long and complex history of federal policies, individual entrepreneurial spirit and creativity, natural disasters, economics, and trade. In part, the future of irrigation depends on what society learns from these experiences.

In the United States, irrigation was first practiced by the indigenous people in the Salt River valley, on the Colorado Plateau, and along many river courses throughout the West. The ancestors of the present-day Pima, Hopi, Tohono O'odham, Hualapai, Havasupai, Yaqui, Pomo, and other American Indians grew corn, peaches, beans, squash, melons, and other crops through an intricate network of ditches and canals. In many instances, today's irrigation canals follow the same general layout of prehistoric canals, such as in the Salt River valley in central Arizona.

Some of the immigrants arriving in North America, particularly from the Mediterranean, brought with them ages-old heritages of watering the land as part of farming. Spanish colonists irrigated extensive gardens at the string of missions established on the Pacific Coast beginning in the 1760s. Spanish water rights are still a part of California water law. However, Northern European colonists had no background in irrigation and found no need for this in the humid East until the population began to push into the arid West.

The history of modern irrigation in the United States can be dated to July 1847, when an advance party of Mormons preceded Brigham Young into Utah's Salt Lake valley and immediately diverted water onto a patch of land to soften the soil so they could plant potatoes. Over time, irrigation became a central component of the federal government's strategy to encourage settlement of the West and build a nation that stretched from coast to coast. Irrigation was extolled as the solution to a range of problems, including overcrowding in eastern cities and debt resulting from the Civil War and earlier wars. Through a series of Executive Orders and Acts of Congress, legislation was enacted that provided for the division of land, the establishment of reservations and allotment and sale of Indian lands, and the settlement and sale of "excess public land." Not by coincidence, one of the major federal water resource development agencies, the U.S. Bureau of Reclamation, was formed during this time, and it provided significant engineering and financial capital to develop irrigation on newly acquired lands in the West.

Even as the push to irrigate was at its peak, however, some people recognized irrigation as a mixed blessing. T. S. Van Dyke (1904) writing in the magazine *Irrigation Age*, put it very well:

> That perversity of human nature that leads us to take hold of so many new subjects by the wrong end seems to rejoice especially in misleading the beginner in irrigation. . . . this perversity may mislead him into thinking he is accomplishing wonders when he is losing money by the day. . . . One may be injuring the land without suspecting it, and about the time he has lost considerable money may conclude that irrigation is a heartless hoax.

Despite the significant contributions that irrigated agriculture makes to society in terms of food and fiber production, society entertains an ambivalent feeling toward irrigated agriculture, with some reason. First, because irrigation is practiced largely in arid regions, although it is increasingly used in subhumid and humid regions to ensure timely availability of water, it is increasingly coming into competition with other sectors of society for a scarce resource—water. Second, although irrigated agriculture surely is not alone in causing negative environmental impacts, irrigation can degrade water quality, deplete streamflows, reduce ground water levels, and alter stream channel morphology and local hydrologic regimes. The nation's sensitivity to such environmental harm has deepened over time.

The remarkable early development of irrigation was fueled by an enormous level of federal involvement, including engineering and financial assistance, but this role has diminished greatly in recent years. Growth no longer needs encouragement. And the role of agriculture in the nation's economy has changed. Only a small percentage of the population now makes its living by producing agricultural products, and memories of the family farm no longer are part of most people's experience. Corporate farms are now common. At the same time, some of the nation's priorities have shifted, with increased concern for environmental issues, an increasingly global economy, and tremendous pressures from population growth and urbanization. Parks, nurseries, urban landscaping, suburban lawns, and golf courses have been added as bona fide users of large quantities of water. Instream uses such as fishing, boating, and preservation of fish and wildlife habitats have greater standing today than they did in the past.

THE COMMITTEE'S CHARGE AND APPROACH

The National Research Council's Committee on the Future of Irrigation in the Face of Competing Demands was formed to study the changing availability of water for irrigation and to identify ways to facilitate irrigation's transition to a world where there are increasing and, in some cases, conflicting needs for water. The committee was asked to draw lessons from past experience, examine current and foreseeable advances in science, and identify examples of change and responses to change that appear to be underway at this time that might tell us something about the future direction of irrigation. Specifically, the committee was asked to

• describe some of the short-term and long-term issues associated with irrigation in both the western and the eastern United States, including, for example, impacts on water quality, soil quality, and environmental values, as well as physical, biological, social, and economic impacts;

• identify the range of pressures affecting the availability of water for irrigation and the impacts of these pressures for major regions of the United States;

• explore the role of technology in helping the nation adapt to changing conditions and identify gaps in the knowledge base; and

• identify and evaluate economic, institutional, and policy changes that might facilitate the transition of irrigation to a world of increasing water scarcity.

The committee used a variety of mechanisms to gather information for this study. The committee's members, selected to represent a diversity of skills, knowledge, and experience, augmented their own knowledge by hosting a workshop with some 40 invited guests, conducting in-depth case studies of different regions in the nation, and visiting Arkansas, California, and Oregon to talk with representatives from agriculture and related fields. In addition, the committee conducted an extensive review of the literature to gain understanding of both the forces of change affecting irrigation and the responses to that change. The committee focused particular attention on the forces behind current changes in the availability of water for irrigation and the types of responses needed to adapt to these different circumstances. Special effort was made to hear from people in different parts of the country, particularly from farmers and others closely involved with agriculture and possessing hands-on experience in water management. Efforts were also made to obtain insights from those representing some of the key competitors for water or concerns for environmental quality. As the committee's work progressed, it became apparent that the report would have a heavy emphasis on the West because that is still where most irrigation occurs—91 percent of all irrigated acreage lies in just 20 states: the 17 states west of the Mississippi River, plus Arkansas, Florida, and Louisiana.

Many ecological, technological, economic, institutional, and social factors affect the future of irrigation. Three broad approaches may be used to group these factors and assess how they might operate in combination with one another over time:

1. Appraise the current situation (i.e., emerging problems, technologies, and policies that raise concerns or offer promise);
2. Develop historical and geographical analogies from past irrigation experience that might have practical value for future decisions; and
3. Examine and extrapolate from recent trends (e.g., in irrigated acreage, production patterns, and problems in the latter part of this century).

This report reflects some of each approach. This chapter introduces the issues and sets the stage for understanding the problems faced. Chapter 2 provides a cultural and historical perspective on how societies adapt to change and how we might best help irrigation evolve in the future. Chapter 3 appraises the status of irrigation today. Chapter 4 highlights the forces of changes that are at work. Chapter 5 provides specific examples of how different states or regions are being affected by change in water availability and cost and the approaches used in response. Chapter 6 draws conclusions from the committee's deliberations.

DEFINING OR DIVINING THE FUTURE?

There is a large body of practical wisdom embedded in North American irrigation experience. The continent has some 2,000 years of irrigation practice, accumulated in a wide range of environmental, cultural, political, and economic contexts. It includes myriad trials, errors, failures, and successes. Some irrigation systems spread over large regions, while others were adapted to local microenvironments and communities. Some lasted centuries; others barely a season or two.

This diversity of experience has several implications. First, it should forewarn us to guard against overgeneralizations about the nature of irrigation systems and their futures. There are always limits to the lessons drawn from analogy. Second, it helps define the range of possible adjustments available to respond to future problems. It is useful to recall the comment of Gilbert Levine (1985):

> Irrigation systems are as much behavioral as technical. They require daily interactions within the irrigation bureaucracies, among farmers, and between the bureaucracies and the farmers. . . . Yet the thought, time, and effort devoted to understanding and dealing with behavioral questions are infinitesimal by comparison to that devoted to the technical issues.

Indeed, one of the biggest limitations on making analogies based on current or past experience lies in our inability to quantify or measure behavioral and cultural factors.

There is also uncertainty inherent in extrapolation that limits what this committee can and cannot say about the future of irrigation. Modern irrigation systems in the United States are less than 100 years old. Some have existed for only a few decades, but during this time they have changed frequently and significantly. Assessing the future thus depends on analyzing many short-term experiments as well as long-term trends. Because long-term data on irrigation exist for relatively small areas and may not be transferable from region to region, it is important to guard against errors made by large-scale regional generalizations based on short-term local evidence. Also, long-term trends, of the time scale of decades to centuries, may shed light on the sustainability of irrigation, but they may also have little immediate relevance for irrigators making decisions on time scales of days, months, seasons, or years. The challenge is to draw analogies from long-term experience that are relevant for short-term decision making (Wescoat, 1991).

This challenge underscores the need to identify relevant time frames and spatial scales. Sustainable development implies a relatively long time period—decades to centuries—during which time short-term problems and crises must also be met. Irrigation planning, by contrast, uses time frames of several years to several decades. Irrigators, in further contrast, may focus on periods of several hours to several years. When addressing the future of irrigation, it is necessary to consider several time frames simultaneously. Short-term decisions may lead toward, or away from, sustainable irrigation development. They may facilitate, or eliminate, long-term water management alternatives. Long-term decisions, such as reservoir construction or indefinite water rights, facilitate some short-term flexibility and options while impeding others. Such decisions may increase vulnerability to low-frequency but high-magnitude natural hazards. It is also important to work at multiple spatial scales—from the household to the globe. Although this report has a national focus, it recognizes that the nation consists of varied regional irrigation patterns, that those patterns vary locally, and that they are influenced by global and international markets and pressures. In any locale, irrigation evolves its own culture, or set of behaviors, in relation to these influences.

Another challenge is our inability to define what constitutes an irrigation "success" or "failure." Success is defined in the present in terms of desired or foreseeable outcomes. In the context of present-day discussions of sustainable development, irrigation success might be defined as the ability to continue farming and to improve the management and productivity of an irrigation system, and at the same time to show resilience to internal or external environmental and economic variability. In earlier decades, however, success was defined more in terms of profitability, uniformity, and economies of scale. Still earlier, subsistence, sufficiency, cooperation, and conflict resolution may have served as the key criteria for successful irrigation.

Times and priorities change, and these changes in turn drive policies and institutions in new directions. Can we judge what will be successful in the future if we do not know what the evaluation criteria will be? We suspect that they will not be the criteria of the past, but in what direction and how far will they change? Divining the future will require defining the criteria for success for the point in time being considered. Defining such criteria will be difficult because there are multiple objectives in play, and the criteria for success will vary among parties with different objectives. Given increasing demand for water and demands from new users, finding balance will be key: balance among users and balance among profit, productivity, and environmental protection.

THE HISTORICAL CONTEXT

To understand the current condition of irrigated agriculture in relation to water supply in the United States, some understanding of the setting in the nineteenth century is essential. As the country expanded westward in the early days of settlement, demand for water was small in relation to supplies, and thus opportunity costs of using water for irrigation were low. Our ability to control flows was very limited—either too much or too little water was an obstacle to the development of about one-third of the area of the original 48 states. Natural supplies were critical factors in shaping the development of the nation: rivers provided the principal paths for exploration and transportation, and consequently cities grew around major rivers and harbors. Agriculture, too, was oriented to water supplies: farmers gravitated to places where precipitation and soil was adequate or where streams could be easily diverted for use on crops. Social pressures exerted influences as well. Federal policies encouraged settlement of semiarid and arid areas, and irrigation became the fundamental cornerstone of the development of the western United States during the latter part of the nineteenth century. To supply an expanding population with food, and to make acquired lands productive, water was harnessed for use in irrigation. While the supply of water in relation to demand was large at this time in most instances, the need to secure a permanent source of water gave rise to water laws, such as the prior appropriation doctrine, which established the principle of "first in time, first in right." The development of the prior appropriation doctrine provided strong incentives for farmers to "use it or lose it" where water was concerned. The prior appropriation doctrine also codified a widely held perception at the time that water left flowing in streams was a "waste." Beneficial use became a condition for securing a water right.

Another important historical element is the relationship between the United States government and Indian tribes. Treaties signed between Indian tribes and the United States in exchange for the former ceding vast territories provided for the reservation of land and an amount of water "sufficient to fulfill the purposes of the reservation" (*United States v. Winters*, 1908). In the latter half of the nineteenth century, the expected use was agriculture, and because irrigation formed the cornerstone of U.S. Indian policy, practicably irrigated acreage became the standard measure of Indian water rights. Although relatively little was done then to secure and develop water supplies for Indian irrigation projects, the dedication of significant quantities of water to tribal use now, as treaties are enacted, has enormous implications for future irrigation in the United States.

At the turn of the century, the welfare and survival of countless people living in arid, semiarid, and flood-prone areas were at risk from year-to-year variation in weather and precipitation patterns. Many irrigators and irrigation projects were hopelessly in debt. Also, with the more easily irrigated lands already developed, the expansion of irrigation depended on storage and conveyance systems to increase dependable water supplies and on technological improvements to pump ground water. Thus a new era began—what might be called the construction era (1900 to 1970)—a period of rapid growth in water use and development of infrastructure. In this period, withdrawals rose from 40.2 to 370 billion gallons per day. Irrigation use of water increased from 20 to 130 billion gallons per day. The number of dams in the nation rose from 3,000 to more than 50,000, and as a result water storage capacity increased from 10 to 753 million acre-feet (Frederick, 1991).

Many factors contributed to this pattern of growth. Technology improved dramatically (e.g., mechanization of earth moving, vertical turbine pumps, large-scale dam construction techniques). Federal policy evolved to reflect the view that it was in the national interest to develop water resources where they were capable of producing crops, power, or other outputs of economic value to the nation. The successes of technology built a national optimism that inspired other achievements. Dams, canals, pumps, and other engineered infrastructure components became the accepted solution to virtually any water problem. Planners were always ready to provide offstream users with virtually unlimited supplies at low cost, with the impacts on streamflows basically ignored. Few people had the vision to give much consideration to long-term impacts.

The significance of the federal role in the development of irrigation should not be underestimated. Federal laws and policies provided the land base, capital, and incentives to settle arid and semiarid western lands. The Desert Lands Act, the Homestead Act, and the Dawes Allotment Act provided the necessary framework for land acquisition and settlement. The Reclamation Service (currently the Bureau of Reclamation) was established specifically for the purpose of "reclaiming" arid and semiarid lands for use in irrigated agriculture and, in general, for the development of the West. Western promoters turned to irrigation as a necessary means to sustain western development. Every reclamation project authorized by Congress between 1902 and 1945 provided that the primary purpose was for development of irrigation.

The construction era ended as the result of many converging factors. Three of the most important factors were the high cost of developing new supplies, the lack of federal funding brought on by increasing federal budget deficits, and environmental and health concerns expressed through public opposition, legal challenges, and environmental legislation.

A look at recent changes in water use, development patterns, and federal policy confirms that there has been a shift in the nation's approach to water use. Per capita water withdrawals peaked in 1975, and total withdrawals peaked in 1980. The construction of water projects peaked in the late 1960s. Related to those trends, federal policies evolved to provide less funding, impose higher interest charges, and require cost sharing for irrigation projects. At the same time, public interest in environmental protection increased, as evidenced in environmental legislation such as the Wild and Scenic Rivers Act of 1968, the National Environmental Policy Act of 1969, the Coastal Zone Management Act of

1972, Federal Water Pollution Control Act Amendments of 1972, and the Endangered Species Act of 1973. The courts and the legislative process became the battleground for major changes in water project development.

Concomitant with these changes came changes in the balance between water supplies and demands. Water demands increased with population growth, increased incomes, increased leisure time and interest in outdoor recreation, and rising environmental values. Supply, on the other hand, was not keeping pace—depletion and degradation of supplies became factors, and the high costs of water treatment and recycling limited the adoption of those supply-enhancing options. Rising water project costs became inevitable because the best reservoir and dam sites were developed first, and the provision of new storage facilities is eventually subject to diminishing returns. Moreover, the opportunity costs of storing and diverting additional water became unacceptable. Overall, the magnitude and nature of future increases in water costs among users will depend in large part on how existing supplies are managed and allocated.

There are clear implications associated with how water is priced. If water is underpriced and supplies are locked into traditional uses, then more of society's costs will take the form of deteriorating aquatic ecosystems, loss of instream values, restrictions on development resulting from the inability to secure adequate supplies, more frequent interruptions in service, and impediments to urban, industrial, and economic growth. In addition, agriculture will continue to consume a large quantity of scarce water in relatively low value uses, and there will be inadequate economic and institutional incentive to adopt more efficient irrigation techniques and strategies.

On the other hand, if the costs of water are borne by users who have incentives to conserve and opportunities to sell water, then there will be benefits to society as water is used more efficiently, the highest-value uses (determined either by market forces or societal goals) are assured of adequate supplies, and the nation derives greater overall net benefits from its water resources. In addition, irrigators would benefit from the increased value of their water rights, would have increased incentives to conserve water or take land out of production, and would have more capital to invest in water conservation technologies.

IRRIGATION: INDUSTRY OR CULTURE?

This study considered many questions, but there was one overriding issue that, while difficult to articulate, seemed recurrent. Agriculture is viewed by the public in two not necessarily consistent ways. The first is that agriculture, including irrigated agriculture, is a business, an industry, albeit an industry essential to human existence. Competing with this pragmatic view is one that sees irrigated agriculture as a complex system, one that has spawned an individual culture.

If society takes the position that irrigated agriculture is an industry, it would logically move toward a situation where the user bears all the costs of production.

In turn, the producer passes those costs along to the customer. On the other hand, if society accepts that irrigation is more a culture—the way people live and part of the national identity—it is then logical for the public to absorb a significant share of the responsibility for the activity in the name of the national interest. Thus society shares in the costs and uncertainties of farming by providing various subsidies to farmers, which in turn subsidize the costs of food and fiber to consumers.

Both models have strengths and weaknesses. Since the 1930s, the United States has favored the agriculture-as-culture model, but the trend in recent years is changing. More and more, agriculture, including irrigated agriculture, is seen as a business that must compete in a global economy. This new emphasis stresses the entrepreneurial side of farming, but it may diminish the cultural assets associated with the irrigated agriculture community. The United States is not alone in this dilemma—it is playing out throughout the world.

This committee met with many people during the course of the study, and it was clear that many of those most active in irrigation and its associated industries view irrigated agriculture as a business. The farmers, in particular, held this view; they are all sophisticated industrialists operating relatively large enterprises, and they see themselves as business people. Even the Bureau of Reclamation, once charged to promote irrigated agriculture as a social goal, now seems to see irrigated agriculture as an industry as well and is seeking to evolve into a new role in water management. From an historical perspective, this is a radical idea. It flies in the face of the nation's history, which soundly supported the subsidizing of irrigated agriculture as critical to the national interest—the settlement and regional development of the West. It is still possible to argue that water is "different" from other commodities and thus should be exempt from the harsh discipline of the market. Although the committee believes that the more successful farmers are moving away from that view, their shift in attitude may not be shared by those engaged in smaller farming operations who struggle to make a living, even with water and crop subsidies.

In considering the cultural issue, a critical dimension is geographic scale. The national scale is an important one to look at with respect to the pressures on irrigated agriculture. These pressures are many and mounting: environmental quality, salinity, urbanization, energy prices, subsidy withdrawal, opportunity costs, and even the uncertainty of climate change—these all play a role in forcing change on the irrigation sector. Among these, the environment has only been a factor of any influence for 25 years, a relatively short period of time. As an issue, environmental concerns emerged from a stage of nonrecognition to a stage of widespread public recognition, and environmental needs are now competing for water with more traditional users. It is difficult to forecast the ultimate impact of environmental concerns on the availability of water for irrigation, but these concerns are likely to remain among the factors to be resolved.

The local scale is the appropriate scale for responding to many types of change. The notion that one national set of policies pertaining to irrigation will achieve desired outcomes in all regions simply is not realistic. For example, Nebraska and Florida face very different problems related to ground water and are developing different strategies to solve them. Beyond the local scale is the individual farm, the location where actual responses to the changing availability and cost of water will take place. Most effective action, and acceptable adaptation, it appears, will take place at the local and farm levels because the problems irrigators face are in the end site specific.

How does this question of irrigation as industry or culture affect policymaking? If the industry view of irrigated agriculture is pursued, one option is to make an even bigger push for markets in water so as to subject the industry to full market discipline. There could be more price pressure on the industry, fewer subsidies, and full-cost pricing of federally supplied water. In return, irrigators might be granted transferrable water rights and limitations on the acreage eligible for federal water might be removed. On the other hand, if irrigation is viewed more as a culture, policy decisions would tend to insulate agriculture from direct market forces. The prevailing view—irrigation as industry or culture—varies from region to region and person to person. Rather than imply that one view is more right or wrong, it is the committee's intention to say simply that both views exist and will continue to exist as irrigation evolves into the future.

THE FUTURE OF IRRIGATION

The history of irrigation in the United States is fundamentally a story about the development of the water resource and the accompanying set of laws, institutions and technologies that have enabled the capture and use of water for irrigation.

The underlying premise of this study is that given the increasing competition for water supplies, changes in how water is managed, allocated, and valued are inevitable. It is not possible to predict the future accurately. Still, change is least disruptive when it is anticipated. Careful thought about the future of irrigation can help the nation as a whole adapt to changing conditions.

This report attempts to identify the range of key trends and factors that are likely to influence the future of irrigation. It addresses competition for water resources, especially from relatively "new" users such as golf courses, lawns, and landscaping, as well as from recreation and instream use. It looks at related issues such as soil and water pollution, federal subsidies for crops and for water, the revolution in biological science, and changes in irrigation technology. It also addresses broad issues with potentially far-reaching impacts such as American Indian water rights, the global agricultural economy, and the changing political climate, each with respect to implications for the future of irrigation.

Irrigated agriculture has played a vital part in the nation's history, and it has served social goals far beyond simply providing food and fiber. Its success spawned a culture and sparked an evolution of technology and myriad supporting institutions. These institutions have served irrigation and the nation well, although not without costs to other water users and social programs. The nation, however, is now facing a time of changing public values and new demands. Irrigators feel a combination of pressures today unlike at any time in the past. They are experiencing competition from new directions, and they are finding ways to adapt. The irrigation sector, like the rest of the economy, is in flux. To succeed in the future, it must be innovative, responsive to change, and a leader in attempts to resolve conflicts with other water users.

The committee cannot say with any degree of certainty how irrigated agriculture will change. It can, however, assert with considerable confidence that irrigated agriculture will change. Few companies produce the same product in the same way they did 50 years ago, and agriculture is no different. Irrigation attained its present stature because it was part of the nation's vision about how best to meet the needs of its citizenry. Whether one agrees or disagrees with the judgments and values that guided past decision makers, it is impossible not to admire the dedication with which that vision was put into action.

REFERENCES

Bajwa, R. S., W. M. Crosswhite, J. E. Hostetler, and O. W. Wright. 1992. Agricultural Irrigation and Water Use. ERS/USDA (Agricultural Information Bulletin No. 638).

Frederick, K. 1991. Water resources: Increasing demand and scarce supplies. In America's Renewable Resources: Historic Trends and Current Challenges. K. Frederick and R. Sedjo, eds. Washington, D.C.: Resources for the Future.

Levine, G. 1985. Irrigation and development. Water Science and Technology Board Newsletter 2(5):1–2. Washington, D.C.: National Research Council.

United States v. Winters 207 U.S. 564 (1908).

Van Dyke, T. S. 1904. Irrigation Age 1:16–17.

Wescoat, J. L., Jr. 1991. Managing the Indus river basin in light of global climate change: Four conceptual approaches. Global Environmental Change: Human and Policy Dimensions 1:381–395.

2

The Cultures of Irrigation

The obvious dimensions of irrigation are tangible—how much water is used, what acreage of land is irrigated, what crops are grown, what forces of change and responses are seen. But to really understand irrigation and how it might evolve in the future, we must consider the more intangible, subjective dimensions of irrigation—in a sense, the context in which change must occur. In this report, we call this the culture of irrigation.

At one level, irrigation is simply the application of water to grow plants. At another level, it is the basis for an economy and a way of life. In a very real sense, irrigation made possible the highly intensive settlement of a landscape that otherwise would not readily support large numbers of people. Irrigation has transformed that landscape, literally and figuratively. The bands of green fields sometimes spreading out to considerable distances from the banks of the rivers of the western United States, the circles of green covering the Great Plains, the urban oases filled with trees, flowers, and lawns—these are the products of irrigation.

More profound than this physical alteration of the landscape is the effect of the human population that accompanied and caused this alteration and whose presence was made possible, in part, because of irrigation. Modern irrigation, beginning in the late nineteenth century, carried with it a sense of mission. People like E. A. Smythe viewed irrigation as nothing less than the progenitor of civilization in an otherwise inhospitable land—the key to making the desert bloom (Smythe, 1905). At this juncture, the roots of modern irrigation have been largely forgotten, although they continue to influence the views of many people associated with western irrigated agriculture, and they help to explain many of the policies and institutions in place today.

There are fundamental cultural dimensions of irrigation. The committee found evidence of these dimensions in its discussions throughout the project, both with those who spoke to the committee and within its own discussions. The committee was surprised, however, at the relative paucity of good research exploring these cultural dimensions, particularly with respect to irrigation in the United States (see Box 2.1).

Culture, as used in this chapter, refers to the "ideas, customs, skills, arts, etc. of a given people in a given period."[1] Irrigation, as it has been practiced in agriculture, is a distinctive activity. It is sufficiently distinctive that it has its own history, its own governmental policies, its own institutions, its own practices, and, historically at least, its own communities. Modern irrigation in the United States probably began with the Mormons whose existence as a community in the Great Salt Basin depended on its practice (Arrington, 1975). It grew in places such as California and Colorado, first in support of mining, and then to support settlement itself. Later in the nineteenth century, it outgrew its utilitarian origins and took on the aura of a movement, becoming for some the basis for building utopian communities (Boyd, 1897), for making the desert bloom (Maass and Anderson, 1978), and for civilizing the Great American Desert of the West (Smythe, 1905). Congress created a federal agency—now called the Bureau of Reclamation—dedicated solely to the task of expanding irrigation in the West (Pisani, 1992).

At the base of this swelling interest in irrigation was a central idea: that a society and an economy could be built on irrigated agriculture. The essentially free, virtually unlimited land area of the western United States could be turned into a productive region, providing land and a means of support for settlers while also producing beef and other agricultural products for the country. It was a bold idea, ideally suited for this era of expansion and exploitation. It found widespread support—not only among those seeking to promote development of the West but also among those in other parts of the country who saw this development as serving their own interests.

Out of these origins grew an irrigation culture that viewed itself as serving a larger national interest as well as providing a means of sustenance in an arid environment. In the arid setting of the West, there was great power in the idea of irrigation. It unleashed remarkable energies of both private and public enterprise in the construction of water collection, diversion, and delivery facilities to make water available for agricultural use. Equally remarkable creativity emerged in the laws and institutions that were developed to support irrigation.

The national-level prominence given to irrigation through the federal reclamation program further supported the development of an irrigation culture. Reclamation projects were extraordinarily successful in obtaining congressional fund-

Box 2.1
Research on Irrigation Cultures

There has been little scientific research on cultural attitudes toward irrigation in the United States. Surprisingly, there has been more social scientific research on irrigation in Africa, Asia, and Latin America than in the United States or Europe.* Cultural research in the United States has tended to focus on prehistoric irrigation and cultural resource inventories near public works projects. The principal exceptions involve ethnographic studies of American Indian and Hispanic irrigation practices and problems (Brown and Ingram, 1987; Castetter and Bell, 1942; Forde, 1963). These patterns suggest that "culture" and "social organization" are associated with faraway places, remote times, and ethnic minorities.

Research in cognate fields of social science has some relevance for irrigation. Environmental hazards research, for example, sheds light on perception and behavioral factors that affect agricultural water use, salinity management, and drought adjustment (Kromm and White, 1992; Riebsame et al., 1991; Saarinen, 1966). Ethnographic research draws attention to the varieties, problems, and importance of local community organization in irrigation management (Coward, 1980; Hunt and Hunt, 1976; Smith, 1972). Sociological research focuses on the dynamics and problems of irrigation bureaucracies at the regional and national levels. Irrigation management research, particularly in Asia, has identified important linkages between cultural, institutional, and technological dimensions of irrigation practice (International Irrigation Management Institute, 1993; Uphoff, 1992). Historical research helps explain the present situation and its problems (Hundley, 1992; Lee, 1980; Pisani, 1992; Tyler, 1992; Worster, 1985).

In addition to these broad fields of academic research, there have been several major syntheses of social and cultural aspects of irrigation. In a comparative study of irrigation districts in the western United States and Spain, Maass and Anderson (1978) combined historical, geographical, economic, and political analyses to assess the performance of local irrigation systems. Detailed comparative analysis of this type is rarely conducted, but it can have great value in irrigation policy analysis. In Indonesia, Lansing (1991) went beyond the conventional opposition between traditional and modern irrigation practices to show how those systems operated in relation to another. He combined ethnographic research with irrigation system modeling to identify solutions to irrigation conflicts and project operation.

American Indian tribal councils are breaking new ground by combining traditional water use norms and practices with modern laws and technologies (e.g., geographic information systems) (Confederated Tribes of the Umatilla Indian Reservation, 1981–1994; Enote, 1995). Some tribes are sharing information with in-

ing support (McCool, 1987). Federally funded irrigation projects sprouted across the West, expanding irrigated acreage in some areas and creating new irrigation in others. Not only did additional lands come under irrigation, but communities developed and grew. Especially in rural areas, these communities often were heavily dependent on irrigation for their existence. Businesses in these communities provided services needed by irrigators, such as the provision of seed, equipment, and basic household supplies. In turn, irrigators generated the market crops that brought outside capital into the community.

Irrigation culture established itself in the quasi-governmental institutions ternational grassroots organizations in developing countries (Enote, 1995). Irrigators and scholars in central Arizona and other regions have studied prehistoric irrigation practices (e.g., canal irrigation, water harvesting, and ethnobotany), in part to guide irrigation into the future (Evenari et al., 1982; Hodge, 1893; Nabhan, 1979). Still, these represent only a handful of cultural studies within the much broader fields of irrigation science.

It seems important to understand why cultural issues have been neglected in irrigation research. This question deserves detailed attention, but four hypotheses may be briefly considered:

1. Cultural perspectives are sometimes viewed as *problems*, rather than as a source of solutions for irrigation problems. For some scientists, culture distorts rational water management. Reformers may regard cultural traditions as obstacles to advances in policy and practice. Culture is often associated with persistent ideological conflicts.

2. Cultural perspectives may seem *unclear* or *contradictory*, even to those who invoke them. The word culture can mean everything, anything, or nothing (Kroeber and Kluckhohn, 1963; Mitchell, 1995; Williams, 1983). Appeals to cultural heritage (i.e., the past) may occur in the same breath with visions of new technology or new policies. Although these complex patterns may in fact represent a contemporary irrigation culture, their utility for policy analysis remains unclear.

3. Cultural perspectives are viewed as *unscientific*, that is, indirect manifestations of more fundamental scientific phenomena that explain irrigation patterns and practices. Culture is symptomatic of, but not the driving force behind, irrigation problems.

4. Finally, culture is often viewed as the *historical* vestige of a bygone era—meaningful to its descendants, but of limited value in the modernizing world. According to this view, cultural factors are viewed as constraints on innovation and adaptation.

Even if partially correct, these views of culture as problem-ridden, vague, unscientific, and anachronistic have not, surprisingly, blocked research on issues that face water managers today and will continue to unfold in the future. It is not likely that specialized research or special interest behaviors will grasp these issues.

---

*This finding applies more to the fields of anthropology, sociology, and geography than economics, law, and behavioral science.

that were established to build and maintain the facilities needed to provide water and in the development of the legal rules for allocating the water itself among different users. Even today, the directors of mutual ditch companies, irrigation districts, and conservancy districts are leading figures in their communities, constituting a power base with considerable influence over water issues at a state and even national level.

In today's increasingly urbanized society, evidence of a culture of irrigation is much less apparent. The unifying idea of a society built around irrigation no longer has the power it once had. It would be a mistake, however, to believe that this traditional irrigation culture is no longer important. It remains alive and well in many parts of the rural West today, and it is also visible in the Great Plains and South. Irrigation continues to support economies in these areas and to make possible a way of living that is otherwise less and less available. It is a culture, however, that in some ways is in retreat and on the defensive. Instead of a national symbol of progress and growth, some now see irrigated agriculture as a depleter and polluter of water, living off government subsidies.

Is there vitality still in the idea of irrigation? If it is no longer to serve as the basis for a society, what is its purpose? What if the "ideas, customs, skill, and arts" of irrigation should be built upon in the future? What must change? These are fundamental considerations in the discussion of the future of irrigation.

This chapter begins by explaining the notion of cultural perspectives in relation to the material presented in the previous chapters. It then discusses five broad cultural themes or issues: understanding the culture of irrigation; cultural heritage within a changing cultural context; cultural diversity; cultural conflict and cooperation; and irrigation knowledge systems.

WHAT ARE CULTURAL PERSPECTIVES AND WHY DO THEY MATTER?

The term "culture" is defined here in four ways:

1. National Irrigation Culture (i.e., widespread irrigation attitudes, perceptions, values, and policies);
2. Local Irrigation Communities (i.e., local community-based attitudes, perceptions, values, and behavior);
3. Complex Processes of Change (i.e., forces and pressures causing change); and
4. Complex Patterns of Change (i.e., responses, often region or site specific, as illustrated later in the regional case studies).

Widespread changes in values, attitudes, norms, aspirations, folklore, and conflicts affect individual and collective decision making, and they have shaped the current situation in irrigation. Among irrigators, these shared attitudes and perceptions have constituted a "culture of irrigation" that has influenced decision making at the national and regional levels. To understand these decisions, and how they might affect the future of irrigation, it is necessary to understand the perceptions and attitudes that shape national and regional irrigation cultures. Understanding local irrigation cultures is often key to resolving conflicts and to identifying and implementing creative practical solutions to irrigation problems.

For example, Chapter 1 suggested that the original social aims of irrigation may have been largely fulfilled, or superseded by other concerns, leading to questions about the need for "a new social contract," "a new era," or a "new vision" for irrigation. Those who perceive these needs are, in a sense, seeking to envision and to design new cultural patterns, contexts, and opportunities for irrigation. Vision and design—these two activities are linked for communities struggling to transform water systems across the country for the twenty-first century.

Of course, not everyone agrees with recent diagnoses of irrigation problems or prognoses for change. Some communities in the western states, for example, argue that the problems are small and easily fixed without radical change. They describe irrigated agriculture as a culture of continuous adjustment and change—hourly, daily, seasonal and long-term adjustments to changing markets, technologies, weather, and social organization—and expect the industry to continue to adjust as needed.

The complex processes of cultural change are clearly evident when looking at specific regions, such as depicted in the case studies in Chapter 5. Thus, a fourth use of the term culture is that associated with the complex patterns found in specific regions. In each case, culture serves as an integrative concept for examining relations among environment, society, and technology; for addressing conflicts; and for expanding the range of alternatives available to future irrigators.

This chapter on cultural perspectives seeks to introduce concepts that recur in later chapters. The "irrigation communities" described in those chapters are the bearers of irrigation culture. The matrix patterns that link irrigation communities with changing technologies, water resources, and markets reflect cultural patterns. The chains of adjustment involve cultural processes. The regional case studies illustrate different cultural as well as economic, technological, environmental, and institutional issues in irrigation. Urban cultures are more prominent in the California and Florida cases than in the Great Plains, where the pattern of ground water irrigation conforms well with rural individualism and small town environments. It should be noted that each sector has multiple subcultures. Whereas some urban residents and developers create landscapes irrigated with vast amounts of potable water, others have banded together to establish new patterns of "xeriscaping," or prairie and desert landscaping.[2] Some of the golf course developments that used vast amounts of irrigation water, fertilizers, and pesticides are advancing to the forefront of the horticultural industry's scientific application of wastewater reuse, wetlands protection, and nonpoint source pollution control. At the same time, water conservation specialists report that some techniques designed to conserve or protect water resources (e.g., lawn sprinkler automation and ditch lining) have unanticipated effects such as loss of incidental vegetation providing wildlife habitat that require further behavioral or technological adjustment.

Irrigated regions have also developed a wide variety of agribusiness and administrative cultures. A key challenge in urbanizing regions is to facilitate multiple and complementary water uses. A challenge in rural areas is to coordinate individual and collective water management at larger and larger scales. American Indian cultures are central in discussions of the future of irrigation in many regions, especially the Pacific Northwest. Different Hispanic and Asian culture groups influence irrigation in California and Florida. Political and legal cultures vary by state, facilitating different kinds of irrigation change and resistance to change. Colorado relies on water courts, while New Mexico and Utah place more responsibility on water administrators. The culture of water as a property right is more highly elaborated—and contested—in the western than in the eastern states. Even the cultures of environmental groups vary across different regions of the country.

Cultural perspectives provide, on the one hand, a synthesis of the diverse factors affecting irrigation decisions—a way of assembling diverse facts, ideas, and insights. They also help identify issues related to the social meaning of, and attitudes toward, irrigation. In both respects, they help us understand current irrigation issues and gauge future possibilities.

CULTURAL ISSUES

The future cannot be predicted from the past (Popper, 1964), and no one situation is exactly like another, but cultural research can help frame analogies to assess the likely strengths and weaknesses of the economic, technological, institutional, and regional alternatives. Analogies use an account of the past (i.e., the analogue) to help imagine, project, or construct a plausible or instructive scenario about the future (Glantz, 1988; Helman, 1988). Irrigators use analogies when they face a problem by reflecting comparable situations in the past, in other regions, or in different resource sectors. Analogies have commonly been used, for example, between the water and electric power sectors. Irrigators use analogies when they imagine how a new technology or crop might affect their operations. Analogies offer a more detailed perspective on contextual factors that influence irrigation than do formal decision models. They can help ensure that all relevant experiences and alternatives are considered. Finally, they are useful for understanding crisis behavior that falls outside the boundary conditions of most irrigation planning and management models. This section uses analogies to examine cultural issues facing water managers today.

Of the many cultural issues surrounding contemporary irrigation, five broad themes stand out. The first of these, "understanding the culture of irrigation," raises basic questions about the nature and meaning of irrigation in the United States. These questions then lead to more specific questions about cultural heritage, context, diversity, conflict, and knowledge.

1. Understanding the Culture of Irrigation. Is there a "culture of irrigation"? That is, do irrigation communities have distinct views of themselves, their contributions to society, attitudes toward water, and institutions they have created? How well understood are these contemporary cultures of irrigation? What types of understanding are needed to resolve emerging water problems and conflicts?

2. Cultural Heritage Within a Changing Cultural Context. What is the "heritage value" of irrigation? How important is it? How does it change as the larger situation changes? What are the options for heritage conservation?

3. Cultural Diversity. What is cultural diversity? Why is it important? What problems does it entail? What are the options for fostering constructive diversity?

4. Cultural Conflict and Cooperation. How has conflict shaped and impeded the development of irrigation? What role have cooperative behaviors played in irrigation development? How do irrigation policies aggravate or alleviate conflict? What are the conditions that facilitate cooperation and conflict resolution?

5. Irrigation Knowledge Systems. What branches of useful irrigation knowledge have been neglected or lost? How might they be identified, evaluated, and adapted? What is needed to support and facilitate innovation and adaptation of irrigation science and practice?

The importance of these questions may be illustrated with examples from modern (1900-1995), early historic (1500-1900), and pre-historic (pre-1500) irrigation (Figure 2.1). These examples also span a range of spatial scales, from the local to the global. They encourage the type of broad long-term thinking needed to ensure the sustainability and timely adaptation of irrigation systems. They also identify alternatives that might be overlooked and keep one mindful of unexpected changes in the context of irrigation (Wescoat, 1984).

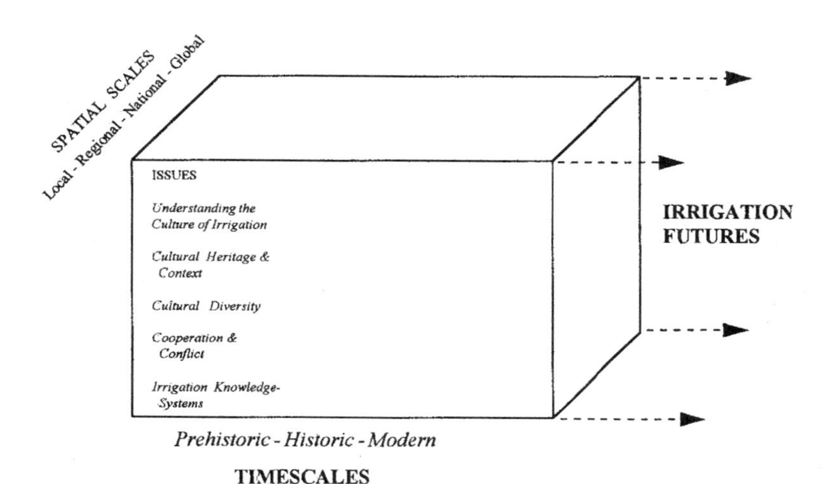

FIGURE 2.1 Conceptual framework illustrating the nonlinear relationships between time, spatial scale, and issues related to the evolution of irrigation.

Understanding the Culture of Irrigation

Is there a culture of irrigation in the United States? There are many distinctive regional patterns and processes of irrigation. The massive agricultural projects and businesses of California, for example, stand in sharp contrast with smaller operations of the Rocky Mountain states. The center pivot systems of the Great Plains have little in common with sugarcane irrigation of the Gulf Coast. There is an enormous diversity of irrigation cultures.

At the same time, several irrigation patterns and movements assumed national significance in the nineteenth and twentieth centuries. Collectively, the perspectives and projects associated with irrigation established a "culture of irrigation," some aspects of which persist, while others face fundamental challenges. Three aspects of this culture seem particularly relevant for the future of irrigation.

The Reclamation Ethic

To this day, many irrigators maintain strong views about the inherent value of "reclamation." Whether draining the bottomlands of the lower Mississippi valley or irrigating the deserts of the West, the historical transformation of "waste into wealth" is a source of enormous satisfaction for irrigators. In arid areas, making "the desert bloom as a rose" has Biblical antecedents and continuing resonance. Indeed, the Reclamation Movement of the late nineteenth century, led by William Smythe (1905) and others, had a missionary zeal and explicitly religious as well as social and economic justifications (Lee, 1980). To participate in the transformation of the deserts and wetlands and to bring out their potential productivity have been viewed as inherently moral and civilizing activities. To settle middle-class families on productive units of land was an inspiring social goal (Mead, 1903, 1920). To meet the harsh challenges of the desert had an heroic quality (Wescoat, 1990).[3] All of these ideas have shaped the view of irrigation as a way of life—and civilization—that has a deep appeal for those who live it.

Many irrigation communities seek to maintain or revive the original values associated with reclamation. They disagree with views that see reclamation as environmentally harmful. Indeed, in some respects the irrigated agricultural community now is paying a price for not responding sooner to early criticism about the harms of reclamation by popular critics such as Reisner (1986). Although many share contemporary concerns for such things as fish, wildlife, and environmental quality, the community overall was badly served by those who initially dismissed the critics. Reclamation agencies adapted slowly and awkwardly to the changing cultural context.

Cast in this light, the ideology of reclamation helps explain some contemporary irrigation problems and conflicts. It calls for a greater measure of respect among the participants in irrigation forums. It also suggests a creative approach to negotiations, which asks, what is the new vision for an integrated water management that includes irrigation in the twenty-first century? What will be the new moral landscapes and new forms of heroism? The history of reclamation informs us that such questions are not irrelevant or utopian: they are practical matters for collaborative work and creative design.

Attitudes Toward Water

In the western states, water often is described as the "lifeblood" of the region. Many in the West still believe that land without water has little value, which is literally true for irrigated cropland. This fundamental dependence on water gave rise to several deeply rooted concepts that guide agricultural water use and have profoundly influenced western water law.

At base, irrigators view water as an essential means to an end—that is, an input needed to grow crops. This highly instrumental view of water promotes the importance of clarity respecting relative rights to use water as well as the value of certainty in those rights. Thus the principle of *priority*—"first in time, first in right"—holds great importance for irrigators. Not only does priority help to sort out competing claims to water, it also serves to protect the substantial claims of irrigated agriculture to water since much of this use was established early enough in the settlement of the West to give agricultural users seniority over most other water uses. One consequence of a priority rule is to emphasize time as the most important factor in determining rights to use water rather than, for example, place, value, or purpose of use (Bates et al., 1993).

Related to this strong desire for certainty is the importance of stability and protection against change. Dependent as they are on the availability of water, irrigators understandably fear the diminishment of their water supply. With a historical record of generally increasing land areas coming under irrigation until the past decade or so, irrigators have jealously guarded their claims. Changes of the use of irrigation water rights, particularly for nonagricultural uses outside the original place of use, have been resisted (MacDonnell and Rice, 1994). The legal concept of *no injury* has emerged to protect the water rights of existing users against change.

At the same time, there is a strongly felt view among most irrigators that water is a shared public resource. This principle is articulated in constitutions and statutes throughout the nation. In addition to serving the interests of individual irrigators, water must be managed wisely to serve the broader interests of the community. This approach is perhaps most completely exemplified by Hispanic *acequia* organizations in northern New Mexico (Crawford, 1989). Riparian principles, prevalent in the eastern states, also emphasize this common property view of water.

The prior appropriation doctrine has never comported with this view of water. It is based on establishing rights to water through the act of capture and use (appropriation), so this doctrine reflects its origins in mining rather than in agriculture. Water rights are regarded as property rights, to be defended as vigorously as any other type of property (if not more so). Water diverted from the river into canals and ditches sometimes has been characterized as private property. That portion of the water supply consumptively used in the growing of crops (perhaps half of the water diverted) is undeniably privatized.

Competing tensions between private and collective need in irrigation produced several important principles. One is the concept of *duty of water.* Legally, this concept places an upper limit on the amount of water necessary to grow crops on a given parcel of land. There is something particularly revealing about the idea that water has a "duty" to grow crops. It emerged primarily as a simple means for state water administrators to allocate water for irrigation uses.[4] It served as a way to more quantitatively articulate the more general principle of *beneficial use*—the condition of water law that the initiation and continuation of a water use are limited to those that are regarded as "beneficial."[5] The doctrine of beneficial use is understood to preclude the *waste* of water (Shupe, 1982). These concepts of duty of water and waste reflect the concerns of irrigators with regards to the importance of water to the larger community that might place limits on private actions. In practice, these principles have rarely been invoked to question established water uses.

Given the increasing competition for water use that often pits irrigation agriculture against urban, tribal, and environmental interests, it is perhaps important to understand these cultural views of water. They help to explain the fervor with which irrigation users sometimes defend their traditional water use prerogatives. They shed light on the resistance of many irrigators to the increased efforts to market water as a means of changing its use from agriculture to cities. They help explain how irrigators may see water as a collective good in relation to the needs of the irrigation community but resist the notion of water as a public, instream resource.

Institution Building

The concepts of priority, beneficial use, duty of water, waste, and injury are formal irrigation institutions as well as attitudes toward water use. Indeed, one of the most remarkable aspects of irrigation culture is the institutions created to guide water use in situations of uncertainty and conflict. In addition to the water rights principles described above, irrigators crafted institutions to administer those rights, such as state engineers, water commissioners, and water masters, whose job was to deliver water within the established water rights structure and to help resolve conflicts among competing water users. Special water courts were created in some states to resolve less tractable water conflicts and to decree the existence of water rights with their priorities. Over the decades, these courts moved from an eclectic set of early precedents to establish impressive bodies of irrigation case law that had far-reaching effects for agricultural and water resources development.

Perhaps the most impressive realms of institution building, however, were those that facilitated cooperation among water users. Irrigation organizations evolved, ranging from local incorporated and unincorporated mutual ditch companies, to larger quasi-public irrigation districts, to multipurpose water conservancy districts (Corbridge, 1984). These organizations served increasing numbers and varieties of users and were able to respond to changing financial, market, and regulatory environments.

Traditional irrigation institutions are increasingly deemed inadequate to meet the challenges of emerging water demands and values, that is, the emerging water cultures. It remains to be seen if irrigation interests can respond to these challenges in modifying existing institutions and creating new ones in a manner that will continue to support irrigation activities.

Cultural Heritage Within a Changing Cultural Context

Two hundred years ago, few American Indians in fishing-based communities would have envisioned the massive depletion of western streams that would ensue because of water development. A hundred years ago, few irrigators would have envisioned water reallocation for stream restoration or policies to conserve and adapt the cultural heritage of irrigation for the coming century. Contemporary pressures on irrigation can only be understood within a broader perspective on local, national, and global change. At every level, and in most regions, the twentieth century has witnessed a shift from agrarian forms of social organization toward various combinations of urban, industrial, environmental, and recreation cultures. Pressure by the latter groups has given rise to concerns about the cultural heritage value of irrigation. Such appeals have much in common with appeals to and debates about the "family farm." It is not clear whether future societies will place more, less, or different values on irrigation agriculture. Cultural change involves the formation of different types of values as well as different weights and relations among values.

Concern about the changing context of agriculture and its implications for public policy are by no means new. The context of federal irrigation policy has varied enormously over the past 150 years, from little involvement in the second half of the nineteenth century, to massive involvement in the mid-twentieth century, and an erratic but generally diminishing role during the past 20 years (Lee, 1980). The 1970s and 1980s witnessed the ascent and maturation of various strands of environmentalism and progressivism, while the 1990s brought sweeping conservatism and dissatisfaction with government programs. A century earlier, the Bureau of Reclamation sought to integrate regional land settlement, resource use, and economic development for small farmers (Mead, 1903). These populist aims did influence reclamation projects in some parts of the Southwest and Rocky Mountain region, but they were alternately supported and rejected, and ultimately redirected toward large agribusiness interests in regions such as California (Pisani, 1984, 1992). For a variety of reasons, the social vision of federal reclamation policy bore little fruit (Lee, 1980).

The first state engineer of California, William Hammond Hall, described the late nineteenth century context of irrigation in the Central Valley as an eclectic melange of groups and practices—the same region that became the most large-scale, centrally planned irrigation system in the country. Earlier, irrigators on the Rio Grande faced dramatic territorial changes with Mexican independence followed by annexation and internationalization by the United States. The first Mormon irrigators in Utah had a relatively stable cultural context, while utopian irrigation communities in northern Colorado and California faced a tumultuous situation from the outset. The first state engineers, like Elwood Mead in Wyoming, helped codify norms and standards of water appropriation and use. Concepts of beneficial use, waste, and the duty of water were designed to respond to variations in cultural as well as environmental conditions.

Federal Indian irrigation policies have, with recent exceptions in the Southwest and Pacific Northwest, been stagnant or regressive for much of this century (Burton, 1991; DuMars et al., 1984; Folk-Williams, 1982; Jacobson, 1989; McCool, 1987; McGuire et al., 1993). In the second half of the nineteenth century, Indian reservations had implicit agricultural purposes, but no clear water rights. The U.S. Supreme Court issued a clear ruling in 1908 that Indian reservations do have water rights reserved for the purposes of the reservation, including irrigation. Despite dramatic change in the legal situation, only about 70 Indian irrigation projects have been authorized, and only a small number of water rights settlements have been enacted. Indeed, the situation has been complicated by efforts by tribes to use their water rights for other purposes, including water marketing and fisheries protection. The latter use recalls an earlier era when the cultural heritage of fishing, although greater than that of irrigation, was ignored by the dominant culture—a lesson for every field and aspect of environmental use and enjoyment.[6]

Even in pre-historic times, changes in the culture occurred. It has been suggested, for instance, that the abandonment of Hohokam canal irrigation involved in-migration by nonagrarian groups that altered the social organization of the region and destabilized large-scale irrigation (Doyel and Plog, 1980). Although other factors were certainly at work, the point is that cultural change alters, for better or worse, the vulnerability, resilience, and adaptability of irrigation systems.

These examples of cultural heritage and change raise basic questions for policy makers. What are the cultural heritage values associated with irrigation agriculture? What difference do heritage values make when land, water, and commodity markets change? These questions are beginning to be addressed in local areas facing water transfer pressures. But even in those places, greater effort is needed to appraise the cultural value of irrigation agriculture relative to other activities within a changing regional, national, and international context. As the cultural context of irrigated regions changes, it will be necessary to document and respond to the changing values associated with irrigation.

When such values are significant, what are the options for ensuring that they are considered when decisions are made? There is a large literature on economic and institutional aspects of heritage conservation, but there have been few systematic applications to irrigation. Which conservation options lead to viable, sustainable irrigation cultures? Which are likely to obstruct beneficial change? Which are likely to yield only "museum pieces," rather than the proper revitalization of some irrigation cultures and retirement of others?

These questions lead to others: What aspects of contemporary irrigation are likely to be valued by future generations? Which actions, taken today, would secure or undermine those values? Expansion of irrigation in humid and urban environments will bring cultural changes and encounter new cultural contexts. What lessons can be drawn from the expansion of irrigation in other regions? What experiments, designed in these new environments, could identify new options for the older irrigated areas of the country? In each case, there is a need to determine which aspects of irrigation endure and which become truly obsolete within a changing cultural context.

Cultural Diversity

Irrigation encompasses an extraordinary diversity of technologies and social forms from pre-historic times to the present. Although frequent reference is made to the great Hohokam canal builders of central Arizona, it is important to recognize that they (like irrigators today) coexisted with many other types of irrigators (Cordell, 1984; Downing and Gibson, 1974). Floodplain gatherers and floodfarmers worked the major, less controllable, channels of the Colorado and Rio Grande rivers (Bryan, 1929, 1941). Runoff harvesting, pebble mulching, and water spreading developed over broader hillslopes and plateaulands (Lightfoot, 1990). Check dams were constructed in narrow valleys of intermittent streams; local springs, wells, and pots brought water to small garden plots (Forde, 1963). Large-scale irrigation canals only developed along managable, mid-sized, perennial streams (Doyel and Plog, 1980; Haury, 1976; McGuire and Schiffer, 1982).

These diverse patterns arose in part through contact with Mesoamerican irrigation centers and partly through indigenous innovation (Doolittle, 1990; Palerm, 1973). Each type of irrigation was shaped by external as well as local pressures and opportunities.

The nineteenth century witnessed the expansion and introduction of new irrigation cultures and the decline of others. Hispanic irrigation, initially established in the late sixteenth century, expanded in the Rio Grande valley, central Texas, southern Arizona, and coastal California (Dobkins, 1959; Hutchins, 1928;

Meyer, 1984; Simmons, 1972). African-Americans introduced and adapted water management and cultivation technologies in the southeastern states (Carney, 1993). Mormons settled central Utah and a constellation of outlying oases (Alexander, 1994; Arrington, 1975). Diverse groups began to irrigate across the western states. English and Scottish farmers established irrigated ranches in Wyoming; Germans in Texas. Italian stonemasons worked on reclamation dams. Chinese laborers reclaimed large areas of the Sacramento River floodplain and San Joaquin delta (Chan, 1986). Other groups arrived in the early twentieth century, including Punjabi and Japanese irrigators in California (Leonard, 1992; Takaki, 1990).

In the late twentieth century, irrigators are increasingly engaged in diversification of farm operations, farm income generation, and manifold varieties of "niche farming" (efforts to develop new combinations of local markets, products, technologies, and agronomic conditions). Large-scale irrigation and ranching operations continue, but they are accompanied by new patterns of urban, recreational, and specialty horticultural irrigation.

The overall picture of irrigation thus continues to be diverse, though some forces serve to increase diversity while others reduce it. Immigration and innovation, for example, increase diversity in their early stages but may ultimately reduce it as existing groups and practices are displaced. In the late nineteenth and early twentieth centuries successful American Indian irrigators such as the Pima of central Arizona were reduced to poverty by upstream diverters (Hackenberg, 1983). Asian irrigators in the western states were first encouraged and then severely persecuted. Mexican workers were hired as laborers for low wages and with little prospect of eventually farming for themselves. Efforts to reduce diversity in the early twentieth century seem to be giving way to celebrations of diversity (e.g., diversity festivals in towns such as Yuma, Arizona) in the late twentieth century.

Diversity has had mixed connotations. Sometimes regarded as an inherent, idealistic quality of American culture, it has at other times been viewed as a problem to be addressed by assimilation policies. In addition to policies aimed at assimilating American Indians and immigrants, governments have imposed large-scale technologies, regulations, and financial policies in areas that supported a wider range of social groups and irrigation practices (Hall, 1886; Hundley, 1992).

The diversity of pre-historic irrigation strategies offers several useful analogies. Ethnobotanists and ethnohistorians have shown that Pima and Papago (Tohono Oodham) irrigators varied their mix of irrigation and food gathering activities in response to climatic variability (Castetter and Bell, 1942; Dobyns, 1974; Hackenberg, 1983; Nabhan, 1979, 1989). Recent research indicates that while some Hohokam groups engaged in intensive canal irrigation, other subgroups used more extensive water management and cultivation methods (Gummerman, 1991). Paleohydraulic research further indicates that many Hohokam canals operated for relatively brief periods of 50 to 100 years, or less, before being replaced by other canals that were less subject to flood hazards, more able to deliver water to large areas, or more advantageous for some villages than others (Howard, 1993). These studies indicate that different culture groups undertook incremental adjustments to flood, drought, desertification, and disaster in ways that yielded a diverse mix of adaptive, and for the most part sustainable, irrigation strategies. Some techniques (e.g., water harvesting, native crops) are being reexamined and adapted for possible future use (Evenari et al., 1982; Nabhan, 1989). Linkages between cultural diversity and economic diversification enabled irrigators to adjust to variable environmental conditions.

Hohokam canal irrigation raises questions about the limits of adjustment in large-scale specialized irrigation systems. Canal irrigators did employ mixed strategies of food production, but they were more tied to a complex, maintenance-intensive, and highly productive system of cultivation than their neighbors. Despite continuous adjustment of irrigation practices, which buffered them against certain types of environmental variability, they were ultimately vulnerable to large-scale systemic collapse.

These themes—diversity, flexibility, and adjustment—changed during the nineteenth and twentieth centuries. On the one hand, the modern irrigators experimented boldly with crops, technologies, institutions, and labor practices (Mead, 1903; Moses, 1986; Robinson, 1977; Smith, 1986; Smythe, 1905). These experiments sometimes involved new social groups, new irrigation practices, and new market niches—thereby creating new irrigation cultures. On the other hand, processes of modern diversification were followed by increasingly rapid diffusion of technology, crops, institutions, and people, which actively serve to reduce diversity to obtain uniform products and economies of scale. Some have maintained themselves for several generations or more, while others have had to seek new markets, and new environmental and cultural niches—or shift to other occupations.

Several conclusions seem relevant for the future of irrigation. First, the history of irrigation is characterized by enormous diversity, which, in principle and in some respects de facto, has been valued by American society. In addition to the inherent value of diversity in a democratic society, it has practical value for risk management, entrepreneurship, and innovation. Because diversity is a dynamic process—not a static relation—it calls for policies that support innovation, risk management, adjustment to local conditions, broad social participation, and access to resources. Access to water is a precondition for cultural and ecological diversity; when denied, it is a source of conflict.

Cultural Conflict and Cooperation

Some of the most inspiring, and painful, lessons of irrigation date to the middle and late nineteenth century, when large-scale population movements displaced indigenous cultures and reworked the water resources of their settlement frontier for mining, farming, and ranching purposes. These processes involved remarkable instances of cooperation, and also bitter conflicts. In addition to conflicts with American Indians over land and water, irrigators had uneven relations with other economic groups. Irrigators established themselves in some areas to serve small populations of miners, travelers, trappers, forts, ranchers, and traders. Where mixed activities flourished, they sometimes came into conflict—as when hydraulic mining destroyed water quality, stream channels, and downstream irrigated lands. Federal reclamation policy sought, in part, to "settle" the western territories, that is, to populate them and to substitute a sedentary stable agrarian economy and society for more volatile and transient activities. Later, when flourishing irrigation economies contributed to population growth and commercial expansion, it sometimes led to competition for limited water supplies and environmental conflict.

Cooperation and conflict are perennial in irrigated areas. Pre-historic irrigation systems involved high levels of community cooperation for construction, maintenance, cultivation, and settlement. The archeological record of economic competition and political conflict is limited, but ethnohistorical evidence suggests that pre-historic irrigators faced a variety of internal and external conflicts that affected the sustainability of their irrigation systems.

In the sixteenth century, Hispanic land and water development extended across central Texas, the Rio Grande valley, southern Arizona, and coastal California, adapting water management practices and institutions derived from Spanish, Roman, and Islamic sources (Baade, 1992; Dobkins, 1959; Ebright, 1979; Glick, 1972; Greenleaf, 1972; Meyer, 1984; Simmons, 1972). Some of these water systems were built with slave labor, while others involved inspiring patterns of community cooperation for canal management combined with private and collective property rights (Hutchins, 1928; Meyer, 1984). Although these rights and the rights of Pueblo Indians were formally recognized in the Treaty of Guadeloupe Hidalgo (1848), conflicts continued to arise over the legal status and protection of community irrigation practices (Brown and Ingram, 1987; Crawford, 1989).

Inspiring lessons of cooperation may also be drawn from the Mormon experience in central Utah and surrounding areas. The Mormon's combination of egalitarian sharing of resources and risks with strong hierarchical decision making helps account for their success in building irrigation systems and communities. Although church control gave way to civil government in Utah rather quickly, it continued to influence community water management in less formal ways (Alexander, 1994; Arrington, 1975). Many small towns in Utah retain the original pattern of large ditches serving fields outside the town and a network of smaller ditches running along streets for family gardens in town (Wescoat, 1990).

Several groups that sought to emulate the Mormon example had mixed results. Utopian and planned agricultural communities like Greeley, Colorado, had high aspirations, but they were not able to attain or maintain the level of community cohesion needed to face internal and external pressures (Boyd, 1897). To be fair in the comparison, they settled in areas of more rapid land development with members of varied backgrounds and goals. But as conflicts arose, these experiments were abandoned for state-based systems of water rights administration.

Most states built on the pluralistic frontier precedents established in miner's courts and people's courts (Moses, 1986). They established ditch companies, irrigation districts, and conservancy districts as civic institutions dedicated to cooperative water development and management. Recently, these groups have been joined by a growing number of nongovernmental alliances of agricultural, environmental, ethnic, and urban groups with interests in local watersheds (Natural Resources Law Center, 1996; Young and Congdon, 1994). It is encouraging to observe progress toward negotiated settlements for issues of longstanding conflict in California, such as, the bay-delta dispute, the Monterey agreement, and water banking and transfer proposals. Equally encouraging is the formation of new groups specializing in alternative dispute resolution and negotiation (Moore, 1995), although progress in this field has not been as rapid as was once expected.

Before turning to the legacy of conflict, it is important to review, again, the extraordinary diversity of nineteenth century irrigation practices. A wide variety of farmers and communities adapted practices from the eastern states for the arid West. African-Americans had transferred rice irrigation practices from Gambia to South Carolina (Carney, 1993). Chinese immigrants played a central role in the reclamation of the Sacramento River floodplains and delta (Chan, 1986). Indians from the Punjab irrigated lands in the Imperial and Central valleys of California, regions very similar to, and influenced by irrigation practices in, colonial India (Jensen, 1988; Leonard, 1992; Wescoat, 1994). American water engineers and lawyers drew practical lessons from Italy and France, as well as India and Egypt (Davidson, 1875; Hilgard, 1886; Jackson et al., 1990; Kinney, 1912; Wilson, 1890-1891). It is little wonder that new types of conflict, and conflict resolution, arose in this rapidly changing heterogeneous environment.

The diversity of irrigation in some areas lacked any coherence. The first California state engineer, William Hammond Hall, wrote despairingly of the Central Valley that:

> Here have met . . . customs of the civil law countries of southern Europe, as modified by Mexican practice; the common law water-course rulings of English courts; and a mining water-right jurisprudence, with customs locally evolved under new conditions. Here also have met, to develop this industry and make laws for its governance, people from all parts of the world and in all grades of circumstances, hardly any of whom had the slightest idea of water-right systems or irrigation customs, legislation, administration or practice. (Hall, 1886, p. 5)

The resolution of this eclectic state of affairs took two paths: cultural conflict, and new institutions to facilitate water development and conflict resolution.

Early in the nineteenth century, it was widely recognized that some groups were violently displacing others. Anglo settlers in the Salt and Gila drainages of Arizona drove Pima irrigators from a profitable cash cropping livelihood to welfare dependence during the last three decades of the nineteenth century (Hackenberg, 1983). Upstream Anglo communities disrupted downstream Anglo irrigation at Greeley, Colorado. Hunters, pastoralists, and mobile farming groups lost access to natural water sources and customary water uses with the expansion of sedentary irrigation and the property rights regimes designed to serve it. As Edward Spicer wrote in 1962, it was the latest "cycle" in three centuries of conquest (Limerick, 1987). Early Asian irrigators who had helped reclaim difficult lands were violently abused, deported, and in some cases stripped of their citizenship.

Water was just one dimension of these social injustices and conflicts. Among the many sobering lessons to be drawn, three seem pertinent here. First, the displacement of some groups by others was sometimes portrayed as natural, necessary, or appropriate—as an instance of the strong and efficient replacing the weak and less productive. Comparable arguments are being addressed toward some irrigators today. How should such arguments be assessed in the light of past experience?

Second, many of the impacts of water development on American Indian tribes were known *as they were occurring*. Although protested by some, they were simply ignored by most (Merritt, 1984). Significantly, the impacts of water development on tribes are continuing. Are there other, comparable, underappreciated, conflicts today. If so, how might they affect the future of irrigation? (Burton, 1991; Jacobson, 1989; McCool, 1987).

Third, the historical literature on irrigation is suffused with high ideals of cooperation and conciliation. Where have such ideals actually guided human action, and where have they been merely rhetorical? It hardly bears repeating that despite Supreme Court recognition of Indian water rights in 1908,[7] and the moral language of treaties, few tribes have obtained their rightful share of material irrigation benefits to date.

In some cases, the adoption of irrigation institutions (e.g., water rights, irrigation district laws, and administrative systems) accelerated the trends mentioned above. In other cases, it promoted inefficient patterns of water ownership and use. But the development of irrigation institutions was in many respects a remarkable achievement with constructive lessons for the future. Irrigation institutions provided principles and procedures for defining rights to water and resolving conflicts among water rights holders. They established organizations to serve collective and increasingly complex social aims, along with rules and regulations to govern those organizations. They asserted that, in contrast with land, water rights are limited and do not include a right to waste.

In future efforts to reform irrigation institutions, three points seem important.

First, irrigation is a cultural as well as economic and political system. The cultural character of irrigation institutions accounts in part for their resistance to change; an understanding of this cultural dimension might facilitate constructive change and conflict resolution. Second, it is important not to lose sight of the enduring value and efficacy of modern examples of cooperation and conflict resolution in irrigation. Otherwise, future generations may find themselves trying to recover what was lost from the middle and late twentieth century as well as from earlier periods. Third, it is important to focus on inventing new forms of cooperation that transcend the costly and historically entrenched patterns of conflict that involve water and related land resources, such as alternative methods of dispute resolution and mediation.

Knowledge Systems in Irrigation: Past, Present, and Future

It is useful to situate the preceding themes within an encompassing theme of irrigation knowledge systems, which speaks directly to issues of policy and research. Knowledge systems include ways of learning and knowing, as well as the types of knowledge obtained, and the relationships between that knowledge and productive human activity. Modern irrigators report rapid adoption of some innovations, such as the annual adoption of new seed varieties, and the (relatively) slow development and diffusion of others, such as the 10 to 15 years for certain subsurface drip irrigation technologies (H. Wuertz, personal communication, 1995). Farmers' "learning curves" vary in accordance with their specific farming situations and pressures, their learning styles, and the relationship between their patterns of communication and decision making. In a rare scientific simulation of the diffusion of irrigation decisions and technologies, Leonard Bowden (1965) found that the diffusion of center pivot irrigation in eastern Colorado could be best predicted by two types of communication—telephone contacts and personal conversations at barbecues!

Some ancient indigenous knowledge systems have endured. The Zuni tribe of New Mexico still draws on traditional agronomic and spiritual practices (Enote, 1995). The tribe is also experimenting with high-tech geographical information systems (GIS) and global positioning systems (GPS), and seeking ways to combine the old and the new. The Confederated Tribes of the Umatilla Indian Reservation in Oregon follow a "tribal philosophy of balance and harmony" to seek cooperative solutions to water conflicts (Hiers, personal communication, 1995). Their *Interim Water Code* (1981, as amended through 1994) balances this tribal philosophy with widespread traditions of western water law. How will tribal knowledge systems be translated into strategies for water use, and what are the implications for irrigated agriculture?

Knowledge systems from the early-modern era also persist, with continuous adaptations and adjustment. Renewed attention is being given to Hispanic *acequia* irrigation and related economic initiatives (Pulido, 1993). The history of irrigation is a lively topic in current debates about the "new western history" (Hundley, 1992; Pisani, 1992; Tyler, 1992; Worster, 1985). Investigations have focused on the historical development and meaning of modern irrigation organizations, projects, leaders, and laws, as well as the more traditional topic of early agricultural settlement. One continuing debate concerns the proper relations among individual irrigators, irrigation organizations, and government agencies. As early as 1874, the reputed environmental observer George Perkins Marsh wrote about the "evils" of irrigation. Maass and Anderson (1978) argued that, contrary to Wittfogel, irrigation organizations had effectively used federal programs to advance their local ends. Donald Worster (1985), by contrast, used Wittfogel's arguments to show how federal agencies colluded with large California agribusinesses to gain control over land, water, and labor in the region. Both arguments have merit for specific areas and events, but neither describes the whole picture.

Another debate concerns the passing of the "water buffaloes" and "lords of yesterday" who built the massive agricultural and urban plumbing systems of the past century. Critics assert that those systems served some groups well in the late nineteenth and early twentieth centuries, but they do not serve the emerging interests of society well at all (Limerick, 1987; Wilkinson, 1988). Irrigation will continue to be part of the cultural heritage in many areas and a vital economic sector in some, but it will no longer be the centerpiece of long-term water planning, use, and power.

Although hotly debated, these interpretations are increasingly influential. At stake are not just the number of acres irrigated, bushels harvested, and acre-feet diverted. There are also serious policy concerns about the economic and environmental performance of irrigation systems, the adjustments needed to fulfill competing demands, and equity issues in water use and reallocation.

CONCLUSION

In question in all of these cultural issues are the lessons, meaning, and value of irrigation. Although it is easy to abuse history, it would be an error to assess current irrigation problems without studying the full record of experience and experiments that created them and that might lead beyond them. Some lessons are inspiring, while others are tragic—many are changing as society rethinks the past as well as the future of irrigation.

Because historical and cultural studies arise from present-day concerns, they shed light on the knowledge systems of the present. They remind us that modern irrigation systems have complex cultural roots, and that they are cultural, as well as economic, institutional, technological, and ecological. In some respects, postwar irrigation science and policy made radical breaks with the past, comparable with the Green Revolution in other parts of the world. These advances in water management and crop production involved ecological and social costs and risks that sparked detailed attention and debate.

More recently, irrigation cultures have been experimenting with new combinations of past, present, and future knowledge—conjoining new technologies and business practices with traditional values, and traditional technologies with new values. These experiments respond to changing local, national, and global situations. They seek to articulate local economic activity with larger regional and global markets. They arise from citizen initiatives and coalitions, supported but not led by government programs and personnel. They seek to balance environmental, social, and economic interests. Finally, they seek more efficient, flexible, and cooperative approaches to conflict resolution.

These experiments have fundamental significance for the future character and sustainability of irrigation agriculture. It is sobering, however, that scientific research on the social aspects of irrigation is advancing further in Asia and other parts of the world than in the United States—a situation that suggests that greater attention be given to comparative international research as well as social investigation of U.S. irrigation planning and policy.

NOTES

1. Webster's New World Dictionary, Second Edition, Simon & Schuster, 1982, at 345.
2. Xeriscaping refers to a range of landscape design concepts that reduce water use.
3. These ideological values should not be exaggerated, relative to the simple economic aims of irrigators, but neither should they be dismissed as merely special interest attitudes.
4. For example, Wyoming under Elwood Mead developed the rule that one cubic foot per second of water diverted during the irrigation season was enough to irrigate 70 acres of land. Later, several other states adopted a rule of apportioning some maximum number of acre-feet of water per acre. For example, North Dakota, South Dakota, and Nebraska place a limit of three acre-feet per acre (Getches, 1984).
5. The doctrine of beneficial use is most fully developed in the western states, while the somewhat broader concept of reasonable use generally applies in the eastern states. Some states use both concepts.
6. Fish protection and multiple use of streams was common in the early legislation of territories like Colorado and was not abandoned for wholesale stream diversion uses until the 1870s.
7. *Winters* v. *United States*, 207 U.S. 564, 28 S.Ct. 207, 52 L.Ed. 340 (1908).

REFERENCES

Alexander, T. G. 1994. Stewardship and enterprise: The LDS Church and the Wasatch Oasis Environment, 1847-1930. Western Historical Quarterly 25:341-364.

Arrington, L. J. 1975. A different mode of life: Irrigation and society in nineteenth century Utah. Agricultural History 49:3-20.

Baade, H. W. 1992. Roman law in the water, mineral, and public land law of the southwestern United States. The American Journal of Comparative Law 40:865-877.

Bates, S., D. Getches, L. MacDonnell, and C. Wilkinson. 1993. Searching Out the Headwaters: Change and Rediscovery in Western Water. Covelo and Washington, D.C.: Island Press.

Bowden, L. 1965. Diffusion of the Decision to Irrigate. Research Paper No. 97. Chicago, Ill.: Department of Geography, University of Chicago.

Boyd, D. 1897. Irrigation near Greeley, Colorado. Washington, D.C.

Brown, F. L., and H. M. Ingram. 1987. Water and Poverty in the Southwest. Tucson, Ariz.: University of Arizona Press.

Bryan, K. 1929. Flood-water farming. Geographical Review 19:444-456.

Bryan, K. 1941. Pre-Columbian agriculture in the Southwest as conditioned by periods of alluviation. Annals of the Association of American Geographers xxxi:219-242.

Burton, L. 1991. American Indian Water Rights and the Limits of the Law. Lawrence, Ks.: University Press of Kansas.

Carney, J. A. 1993. From hands to tillers: African expertise in the South Carolina rice economy. Agricultural History 67:1.

Castetter, E. F., and W. H. Bell. 1942. Pima and Papago Agriculture. Inter-American Studies No. 1. Albuquerque, N. Mex.: University of New Mexico.

Chan, S. C. 1986. This Bittersweet Soil: The Chinese in California Agriculture, 1860-1910. Berkeley, Calif.: University of California Press.

Confederated Tribes of the Umatilla Indian Reservation. 1981-1994. Interim Water Code. Pendleton, Ore.

Corbridge, J. N., Jr., ed. 1984. Special Water Districts: Challenge for the Future. Boulder, Colo.: Natural Resources Law Center.

Cordell, L. S. 1984. Prehistory of the Southwest. New York: Academic.

Coward, E. W., Jr., ed. 1980. Irrigation and Agricultural Development in Asia: Perspectives from the Social Sciences. Ithaca, N.Y.: Cornell University Press.

Crawford, S. 1989. Mayordomo: Chronicle of an Acequia in Northern New Mexico. New York: Anchor Books.

Davidson, G. 1875. Irrigation and Reclamation of Land for Agricultural Purposes in India, Egypt, Italy, etc. Executive Doc. No. 94. U.S. Senate, 44th Cong. Washington, D.C.

Dobkins, B. 1959. The Spanish Element in Texas Water Law. Austin, Tex.: University of Texas Press.

Dobyns, H. F. 1974. The Kohatk: Oasis and Akchin horticulturalists. Ethnohistory 21:317-327.

Doolittle, W. E. 1990. Canal Irrigation in Prehistoric Mexico: The Sequence of Technological Change. Austin, Tex.: University of Texas Press.

Downing, T., and M. Gibson. 1974. Irrigation's Impact on Society. Anthropological Paper No. 25. Tucson, Ariz.: University of Arizona.

Doyel, D., and F. Plog, eds. 1980. Current Issues in Hohokam Prehistory. Anthropological Research Papers No. 23. Tempe, Ariz.: Arizona State University.

DuMars, C. T., M. O'Leary, and A. E. Utton. 1984. Pueblo Indian Water Rights: Struggle for a Precious Resource. Tucson, Ariz.: University of Arizona Press.

Ebright, M. 1979. Manuel Martinez's Ditch Dispute: A Study in Mexican Period Custom and Justice. New Mexico Historical Review 54: 21-34.

Enote, J. 1995. Conservation at Zuni Pueblo: Lessons in Sustainability. Conference on Sustainable Use of the West's Water. Boulder, Colo.: Natural Resources Law Center.

Evenari, M., et al. 1982. The Negev: The Challenge of a Desert. Cambridge, Mass.: Harvard University Press.

Folk-Williams, J. A. 1982. What Indian Water Means to the West: A Sourcebook. Santa Fe, N. Mex.: Western Network.

Forde, C. D. 1963. The Hopi and Yuma: Flood farmers in the North American desert. In Habitat, Economy and Society: A Geographical Introduction to Ethnology. New York: E.P. Dutton. Pp. 220-259.

Getches, D. 1984. Water Law in a Nutshell. St. Paul, Minn.: West Publishing Co.

Glantz, M., ed. 1988. Societal Responses to Regional Climate Change: Forecasting by Analogy. Boulder, Colo.: Westview Press.

Glick, T. F. 1972. The Old World Background of the Irrigation System of San Antonio, Texas. Southwestern Studies No. 35. El Paso, Tex.: University of Texas at El Paso.

Greenleaf, R. E. 1972. Land and water in Mexico and New Mexico, 1700-1821. New Mexico Historical Review 47:85-112.

Gummerman, G. J., ed. 1991. Exploring the Hohokam: Prehistoric Desert Peoples of the American Southwest. Albuquerque, N. Mex.: University of New Mexico Press.

Hackenberg, R. A. 1983. Pima and Papago ecological adaptations. In Handbook of North American Indians, Vol. 10. Washington, D.C.: Smithsonian Institution. Pp. 161-177.

Hall, W. H. 1886. Irrigation Development. Sacramento, Calif.: State Office.

Haury, E. W. 1976. The Hohokam, Desert Farmers and Craftsmen: Excavations at Snaketown, 1964-1965. Tucson, Ariz.: University of Arizona Press.

Helman, D. A. 1988. Analogical Reasoning: Perspectives of Artificial Intelligence, Cognitive Science, and Philosophy. Dordrecht: Kluwer.

Hilgard, E. W. 1886. Alkali Lands. University of California, College of Agriculture, App. 7. Sacramento: State Office, J.J. Ayers, Supt. State Printing.

Hodge, F. W. 1893. Prehistoric irrigation in Arizona. American Anthropologist VI:323-330.

Howard, J. B. 1993. A paleohydraulic approach to examining agricultural intensification in Hohokam irrigation systems. In Research in Economic Anthropology, Supplement 7, V. L. Scarborough and B. L. Isaac, eds. Greenwich: JAI Press, Inc. Pp. 263-324.

Hundley, N., Jr. 1992. The Great Thirst: Californians and Water, 1770's-1990's. Berkeley, Calif.: University of California Press.

Hunt, R., and E. Hunt. 1976. Canal irrigation and local social organization. Current Anthropology 17:389-411.

Hutchins, W. A. 1928. The community acequia: Its origin and development. Southwestern Historical Quarterly xxxi:261-284.

International Irrigation Management Institute (IIMI). 1993. Annual Report. Columbo, Sri Lanka: IIMI.

Jackson, W. T., R. F. Herbert, and S. R. Wee, eds. 1990. Engineers and Irrigation: Report of the Board of Commissioners on the Irrigation of the San Joaquin, Tulare, and Sacramento Valleys of the State of California, 1873. (Reprinted as Engineering Historical Studies No. 5. Fort Belvoir: Office of History, U.S. Army Corps of Engineers.)

Jacobson, J. E. 1989. A promise made: The Navaho Indian Irrigation Project and water politics in the American West. Ph.D. dissertation, Department of Geography, University of Colorado, Boulder.

Jensen, J. M. 1988. Passages from India: Asian Indian Immigrants in North America. New Haven, Conn.: Yale University Press.

Kinney, C. S. 1912. A Treatise on the Law of Irrigation and Water Rights etc., 2nd ed., 4 vols. San Francisco, Calif.: Bender-Moss Company.

Kroeber, A., and C. Kluckhohn. 1963. Culture: A Conceptual Review of Concepts and Definitions. New York: Vintage.

Kromm, D. E., and S. E. White, eds. 1992. Groundwater Exploitation in the High Plains. Lawrence, Ks.: University Press of Kansas.

Lacey, M. J., and M. O. Furner, eds. 1993. The State and Social Investigation in Britain and the United States. Cambridge, Mass.: Cambridge University Press.

Lansing, J. S. 1991. Priests and Programmers: Technologies of Power in the Engineered Landscapes of Bali. Princeton, N.J.: Princeton University Press.

Lee, L. B. 1980. Reclaiming the American West: An Historiography and Guide. Santa Barbara, Calif.: ABC-Clio.

Leonard, K. I. 1992. Making Ethnic Choices: California's Punjabi Mexican Americans. Philadelphia, Pa.: Temple University Press.

Lightfoot, D. 1990. The prehistoric pebble-mulched fields of the Galisteo Anasazi: Agricultural innovation and adaptation to environment. Ph.D. dissertation, Department of Geography, University of Colorado, Boulder.

Limerick, P. N. 1987. The Legacy of Conquest: The Unbroken Past of the American West. New York: W.W. Norton and Co.

Maass, A., and R. L. Anderson. 1978. . . . and the Desert Shall Rejoice: Conflict, Growth and Justice in Arid Environments. Cambridge, Mass.: MIT Press.
MacDonnell, L., and T. Rice. 1994. Moving agricultural water to cities: The search for smarter approaches. Hastings West-Northwest Journal 2:27-54.
Marsh, G. P. 1874. Irrigation: Its evils, the remedies, and the compensation. 43 Cong. 1 sess., S. Misc. Doc. 55.
McCool, D. 1987. Command of the Waters: Iron Triangles, Federal Water Development, and Indian Water. Berkeley, Calif.: University of California Press.
McGuire, R. H., and M. B. Schiffer. 1982. Hohokam and Patayan: Prehistory of Southwestern Arizona. New York: Academic Press.
McGuire, T. R., W. B. Lord, and M. G. Wallace, eds. 1993. Indian Water in the New West. Tucson, Ariz.: University of Arizona Press.
Mead, E. 1903. Irrigation Institutions. New York: Macmillan Company.
Mead, E. 1920. Helping Men Own Farms. New York: Macmillan Company.
Merritt, R. H., 1984. The Corps, the Environment and the Upper Mississippi River Basin. Historical Division, Office of Administrative Services, Office of the Chief of Engineers. Washington, D.C.
Meyer, M. C. 1984. Water in the Hispanic Southwest: A Social and Legal History, 1550-1850. Tucson, Ariz.: University of Arizona Press.
Mitchell, D. 1995. There's no such thing as culture: Towards a reconceptualization of the idea of culture in geography. Transactions of the Institute of British Geographers New Series 20:102-116.
Moore, L. 1995. Regional water planning in New Mexico: An opportunity for citizen involvement in state government. In Sustainable Use of the West's Water. Boulder, Colo.: Natural Resources Law Center.
Moses, R. J. 1986. The Historical Development of Colorado Water Law. In Tradition, Innovation and Conflict: Perspectives on Colorado Water Law. L. J. MacDonnell, ed. Boulder, Colo.: University of Colorado, Natural Resources Law Center. Pp. 25-40.
Nabhan, G. P. 1979. The ecology of floodwater farming in arid southwestern North America. Agro-Ecosystems 5:245-255.
Nabhan, G. P. 1989. Enduring Seeds: Native American Agriculture and Wild Plant Conservation. San Francisco, Calif.: North Point Press.
Natural Resources Law Center. 1996. The Watershed Sourcebook: Watershed-Based Solutions to Natural Resources Problems. Boulder, Colo.: Natural Resources Law Center.
Palerm, A. 1973. Obras Hidraulicas prehispanicus, en el sistema lacustre del valle de Mexico. Mexico City: Instituto Nacional de Antropologia e Historia.
Pisani, D. J. 1984. From the Family Farm to Agribusiness: The Irrigation Crusade in California and the West, 1850-1931. Berkeley, Calif.: University of California Press.
Pisani, D. J. 1992. To Reclaim a Divided West: Water, Law and Public Policy, 1848-1902. Albuquerque, N. Mex.: University of New Mexico Press.
Popper, K. 1964. The Poverty of Historicism. New York: Harper Torchbooks.
Pulido, L. 1993. Sustainable development at Ganados del Valle. In Confronting Environmental Racism: Voices from the Grassroots. R. D. Bullard, ed. Boston: South End Press. Pp. 123-140.
Reisner, M. 1986. Cadillac Desert: The American West and Its Disappearing Water. New York: Penguin Books.
Riebsame, W. E., S. Changnon, and T. R. Karl. 1991. Drought and Natural Resource Management in the U.S.: Impacts and Implications of the 1987-89 Drought. Boulder, Colo.: Westview Press.
Robinson, M. C. 1977. Water for the West: The Bureau of Reclamation, 1902-1977. Chicago, Ill.: Public Works Historical Society.
Saarinen, T. F. 1966. Perception of the Drought Hazard on the Great Plains. Research Paper No. 106. Chicago, Ill.: Department of Geography, University of Chicago.

Shupe, S. 1982. Waste in western water: A blueprint for change. Oregon Law Review 61:483.

Simmons, M. 1972. Spanish Irrigation Practices in New Mexico. New Mexico Historical Review 47:135-150.

Smith, C. L. 1972. The Salt River Project: A Case Study in Cultural Adaptation to an Urbanizing Community. Tucson, Ariz.: University of Arizona Press.

Smith, K. L. 1986. The Magnificent Experiment: Building the Salt River Reclamation Project 1890-1917. Tucson, Ariz.: University of Arizona Press.

Smythe, E. A. 1905. The Conquest of Arid America, 2nd ed. Reprint. Seattle, Wash.: University of Washington Press.

Spicer, E. H. 1962. Cycles of Conquest: The Impact of Spain, Mexico and the United States on Indians of the Southwest. Tucson, Ariz.: University of Arizona Press.

Steward, J. H., R. M. Adams, D. Collier, A. Palerm, K. A. Wittfogel, and R. L. Beals. 1955. Irrigation Civilizations: A Comparative Study. Social Science Monographs No. 1. Washington, D.C.: Pan American Union.

Takaki, R. 1990. Strangers from a Different Shore: A History of Asian-Americans. New York: Penguin.

Tyler, D. 1992. The Last Water Hole: The Colorado–Big Thompson Project and the Northern Colorado Water Conservancy District. Niwot, Colo.: University Press of Colorado.

Uphoff, N. T. 1992. Learning from Gal Oya: Possibilities for Participatory Development and Post-Newtonian Social Science. Ithaca, N.Y.: Cornell University Press.

Wescoat, J. L., Jr. 1984. Long-term change in water management systems. In Transactions, International Commission on Irrigation and Drainage. New Delhi: International Commission on Irrigation and Drainage.

Wescoat, J. L., Jr. 1990. Challenging the Desert. In The Making of the American Landscape. M. P. Conzen, ed. Boston, Mass.: Unwin Hyman. Pp. 186-203.

Wescoat, J. L., Jr. 1994. Water rights in South Asia and the United States: Comparative perspectives, 1873-1993. Paper presented to SSRC Study of Comparative Property Rights.

Wilkinson, C. F. 1988. To settle a new land: An historical essay on water law in Colorado and in the American West. In Water and the American West: Essays in Honor of Raphael J. Moses. D. H. Getches, ed. Boulder, Colo.: Natural Resources Law Center, University of Colorado. Pp. 1-18.

Williams, R. 1983. Keywords: A Vocabulary of Culture and Society. London: Fontana.

Wilson, H. M. 1890-1891. Irrigation in India. In 12th Annual Report, U.S. Geological Survey, Part II, Irrigation. Washington, D.C.: U.S. Geological Survey.

Worster, D. 1985. Rivers of Empire: Water, Aridity and the Growth of the American West. New York: Pantheon Books.

Young, T., and C. Congdon. 1994. Plowing New Ground: Using Economic Incentives to Control Water Pollution from Agriculture. Oakland, Calif.: Environmental Defense Fund.

3

Irrigation Today

Less than 1 percent of the nation's farmland was irrigated in 1900, but by 1982 irrigation accounted for 1 of every 8 acres under cultivation and nearly $4 of every $10 of the value of crop production (U.S. Department of Agriculture, 1986). This transition was driven by economic change: in the late nineteenth century, western promoters turned to irrigation when mining, open range cattle, and dry farming economies proved unable to sustain western settlement (Webb, 1931). During this period, foundations were laid to support irrigation—water rights laws, advances in engineering, mutual water district organization and financing—and these supported early irrigation in areas such as California, Colorado, and Utah. However, irrigation did not begin to expand rapidly until after Congress passed the Reclamation Act of 1902, which established the Reclamation Service (now the Bureau of Reclamation) to assist in developing the West through irrigation.

The federal role in water development expanded further in the 1930s as water development was also used to create new jobs. By the end of World War II, four federal agencies—the U.S. Army Corps of Engineers, the Bureau of Reclamation, the Tennessee Valley Authority, and the Soil Conservation Service (now the Natural Resource Conservation Service of the Department of Agriculture)—

NOTE: There are many sources of data that describe the status of irrigation in the United States. However, the methods used to gather and interpret statistics vary significantly, resulting in disparities among the different sources. Because many references and sources were used in developing this chapter, there are occasions where values may not be fully compatible.

had expanded their roles in the use and development of water resources (National Research Council, 1992a). After World War II, irrigated agriculture expanded rapidly in the far West and the central Great Plains. More recently, supplemental irrigation has become important in the East, Southeast, and Midwest. Irrigated agriculture remained an engine of western development until the 1970s. However, increasing development costs, reduced government financing, increasing demand for municipal and industrial water supplies, diminishing sources of water supply, and a growing concern for the environment have forced water managers and planners to begin rethinking traditional approaches to water management (National Research Council, 1992b).

This chapter provides background information about the current status of irrigation—the amount of land irrigated, types of crops, water withdrawals, and consumptive use. It gives an overview of the technologies used and the economics of irrigated systems, including water pricing and marketing. It highlights key issues in the relationship of irrigation to the environment and introduces an increasingly important force in the water arena: the turfgrass sector. It also highlights another element certain to be key in the future—irrigation on Indian lands. Together, these discussions are designed to provide a quick review of irrigation today and thus set the stage for the committee's foray into irrigation's future. Readers already well-versed in the status and trends of irrigation today are encouraged to proceed to Chapter 4, where the committee explores the deeper cause and effect relationships that underlie the statistics.

IRRIGATED AGRICULTURE

Irrigated Land in Farms

Irrigated agriculture occurs on just 14.8 percent of the harvested cropland and yet produces 37.8 percent of the value of crops (Figure 3.1). The relatively

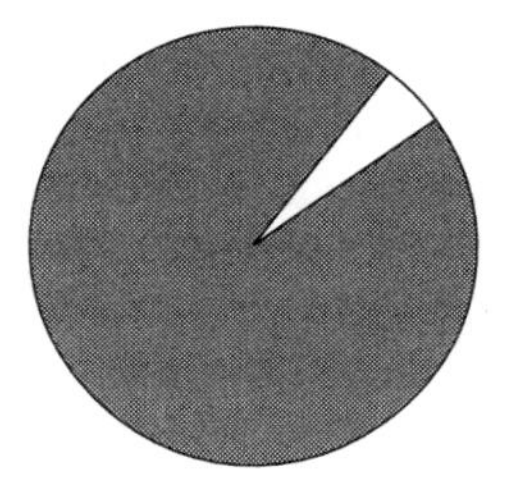

Irrigation accounts for 4.8% of the land in farms (total farm acreage = 964 million acres)

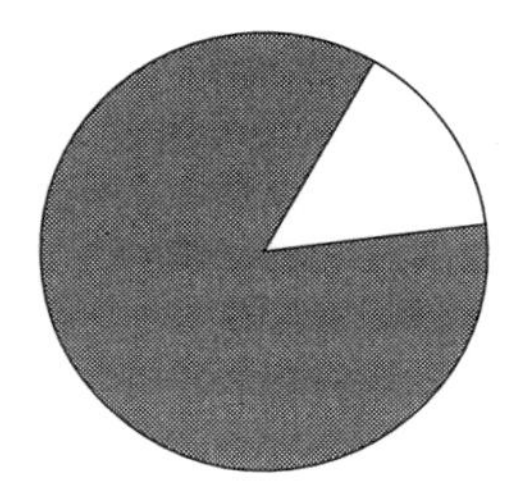

Irrigation accounts for 14.8% of the total harvested cropland (282 million acres)

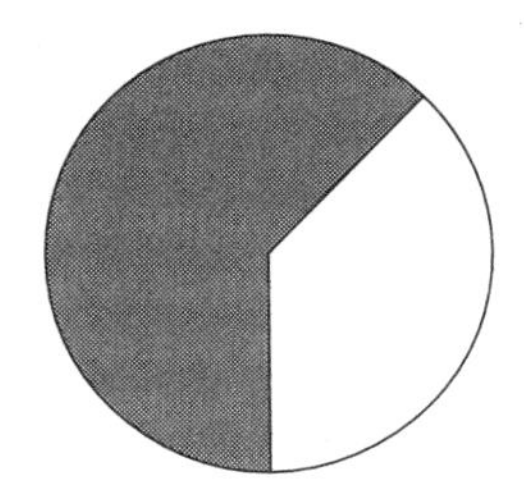

Irrigation accounts for 37.8% of the total crop value ($68.8 billion)

FIGURE 3.1 Irrigation and farm production (*1987 Census of Agriculture*). Source: Bajwa et al., 1992.

Box 3.1 The Top 20 Irrigation States

Twenty states accounted for 97 percent of U.S. irrigation withdrawals in 1995. California and Idaho were by far the largest users of irrigation water and together accounted for 34 percent of the national total. Combined irrigation withdrawals in the four largest withdrawal states—California, Idaho, Colorado, and Montana—exceeded 75 million acre-feet, or nearly half of total U.S. irrigation withdrawals. Florida withdrew the most water for irrigation in the East, although it ranked thirteenth nationwide.

large economic contribution of irrigated agriculture can be explained by the higher yields obtained for irrigated crops, the tendency to irrigate high-valued and/or specialty crops, and the improved product quality and consistency.

In 1959, 9 percent of all farms reported some irrigated land. By 1987, that share had risen to 14 percent. During the 1980s, the total number of irrigated acres and irrigated acres per farm fluctuated considerably because of the temporary idling of land associated with annual commodity program acreage restrictions. Most (90 percent) of the nation's irrigated land is harvested cropland, but many of the mountain states irrigate pasture and land from which wild hay is cut to sustain livestock through the winter.

In the United States, irrigation is used mainly in the 17 western states, plus Arkansas, Florida, and Louisiana (see Box 3.1). These 20 states account for 91 percent of all U.S. irrigated acreage and 82 percent of all irrigated farms. The 17 western states alone contain over 81 percent of the total irrigated land; 85 to 90 percent of total water withdrawn in the West is used for irrigation. Although irrigated cropland provided a substantial portion of national farm income in 1987, there were only about 292,000 individual irrigators, 14 percent of all farmers. Four-fifths of the irrigators were located in the 17 western states.

The drought years of the 1950s and the development of centrifugal pumps and more economical power sources stimulated irrigation development in the southern Great Plains, where ground water is pumped from the Ogallala aquifer. With the advent of the center pivot sprinkler irrigation systems, and with ground water readily available, irrigation expanded rapidly in the central Great Plains during the 1960s and 1970s. Irrigation also expanded in humid southeastern states as a way to provide dependable and timely water. In California and the Pacific Northwest, irrigated areas also expanded during the 1950s and 1960s as many irrigation projects constructed by the Bureau of Reclamation and local entities were completed and put into service. The total irrigated area essentially stabilized in the 1980s due to a combination of low farm commodity prices, increased energy costs, and declining water resources. The percent of harvested cropland irrigated by state is given in Figure 3.2 (Bajwa et al., 1992).

Figure 3.3 shows trends for irrigated land in farms and water applied per acre from share of total acreage irrigated are rice (100 percent), orchards (81 percent),

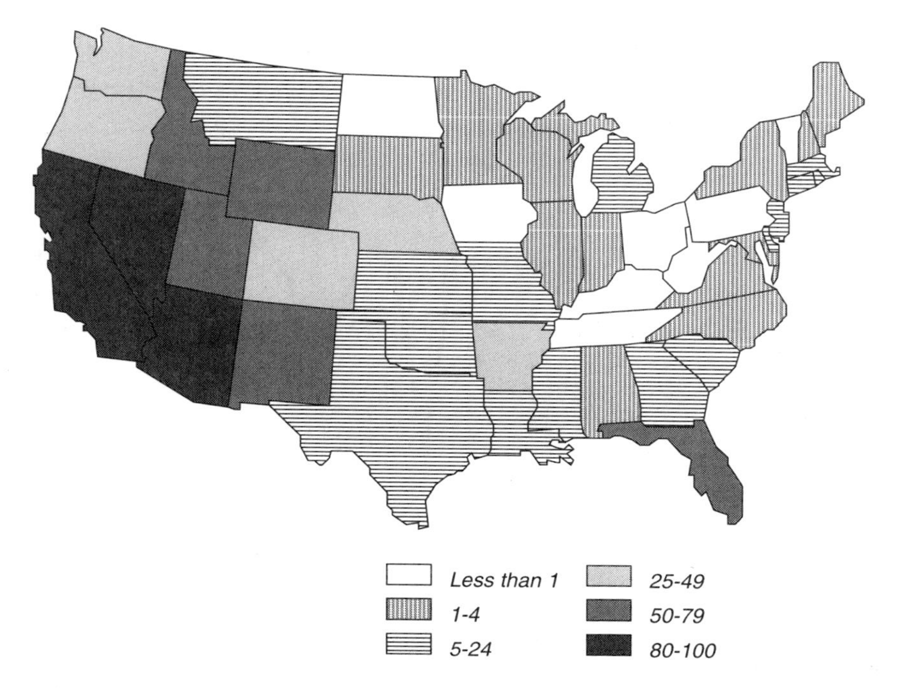

FIGURE 3.2 Percentage of harvested cropland irrigated by state, 1987. Source: Bajwa et al., 1992.

vegetables (64 percent), and cotton (38 percent). Crops with the largest total irrigated acreages are hay, corn for grain, wheat, and cotton (U.S. Department of Agriculture, 1986). While much smaller percentages of grain acreages are irrigated (e.g., 14 percent of harvested corn, 13 percent of sorghum, and 7 percent of wheat), the combination of improved yields on irrigated farms and the increase in the relative acreage devoted to irrigation accounted for 28 percent of the national increase in corn production, 20 percent for sorghum, and 12 percent for wheat from 1950 to 1977 (Frederick and Hanson, 1982).

WATER USE FOR IRRIGATION[1]

Irrigation water typically is measured in terms of withdrawals or consumptive use. Withdrawals represent the amount of water diverted from a surface source or removed from the ground. Consumptive use is a measure of water lost to the immediate water environment through evaporation, plant transpiration, incorporation in products or crops, or consumption by humans and livestock. Consumptive use in agriculture is primarily crop evapotranspiration, which is influenced heavily by climate and the types of crops irrigated. Reasonably accu-

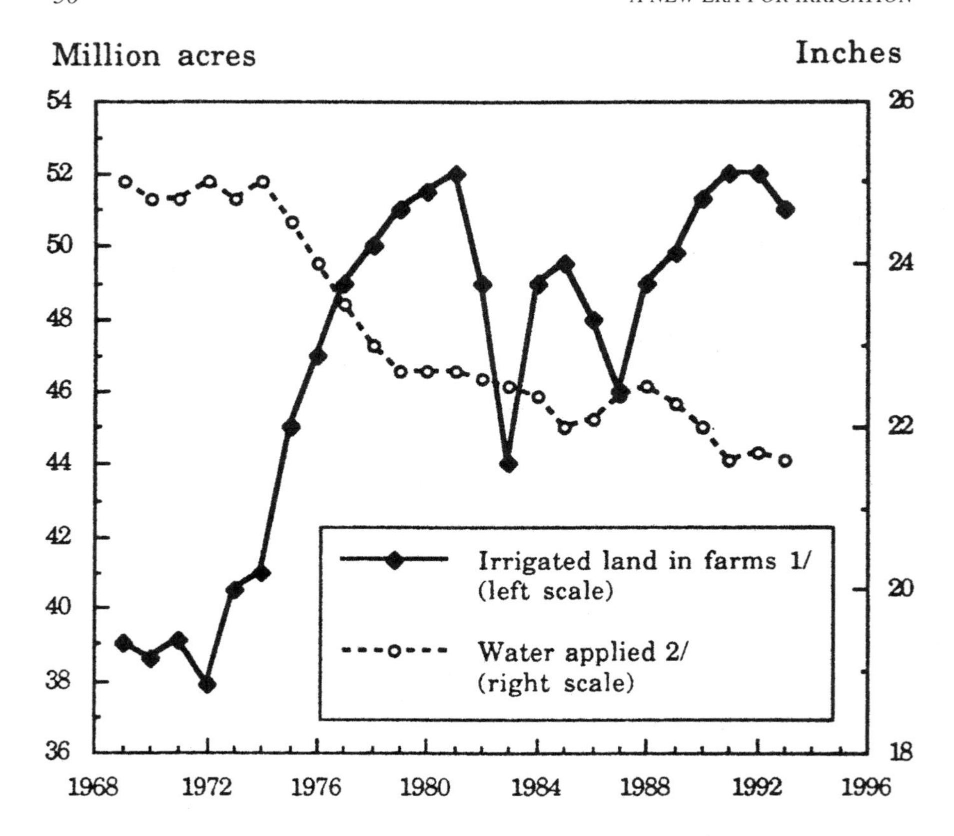

FIGURE 3.3 Trends in irrigated acres between 1969 and 1993. Source: U.S. Department of Agriculture, 1993.

rate estimates of water withdrawn for irrigation can be made if the acreage irrigated, water application rates, and conveyance losses are known. However, reliable estimates for consumptive use and conveyance loss are not currently available. Thus the available estimates are rough approximations of actual conditions. These estimates reflect the importance of the four influential factors: irrigation technology, crop prices, annual commodity program acreage restrictions, and weather. Relaxed acreage restrictions, improved irrigation technology, and high crop prices in the 1970s accelerated irrigation development, increasing total irrigated area from 38 million acres in 1972 to 52 million acres in 1981. Irrigated

TABLE 3.1 Irrigated Area in the United States

Region	1987 (thousand acres)	1992 (thousand acres)	Change (%)
Alabama	84	82	-2
Arizona	914	956	5
Arkansas	2,406	2,702	12
California	7,596	7,571	0
Colorado	3,016	3,170	5
Connecticut	7	6	-19
Delaware	61	62	2
Florida	1,623	1,783	10
Idaho	3,219	3,260	1
Illinois	208	328	58
Indiana	170	241	42
Iowa	92	116	25
Kansas	2,463	2,680	9
Kentucky	38	28	-27
Maine	6	10	69
Maryland	51	57	12
Massachusetts	20	20	-1
Michigan	315	368	17
Minnesota	354	370	5
Mississippi	637	883	39
Missouri	535	709	33
Montana	1,997	1,976	-1
Nebraska	5,682	6,312	11
Nevada	779	556	-29
New Hampshire	3	2	-41
New Jersey	91	80	-12
New York	51	47	-8
North Carolina	138	113	-18
North Dakota	168	187	11
Ohio	32	29	-9
Oregon	1,648	1,622	-2
Pennsylvania	30	23	-22
Rhode Island	3	3	-15
South Dakota	362	371	3
Tennessee	38	37	-2
Texas	4,271	4,912	15
Utah	1,161	1,143	-2
Vermont	2	2	16
Virginia	79	62	-22
Washington	1,519	1,641	8
West Virginia	3	3	-12
Wisconsin	285	331	16
Wyoming	1,518	1,465	-4
Total (43 states)	43,671	46,319	6.1

Source: U.S. Department of Commerce, 1994.

acres then dipped from 1983 to 1987, primarily as a result of acreage restrictions in commodity programs. Water applied per acre has declined from about 25 inches to less than 22 inches.

According to the *1992 Census of Agriculture* (U.S. Department of Commerce, 1994) the total 1992 irrigated area was 46.3 million acres, up 2.6 million acres from 1987 (Table 3.1). During this period, there was no increase in irrigated acres in the West from surface water. The increase in irrigation from 1987 to 1992 occurred mostly in the Great Plains region, which relies primarily on ground water. On the other hand, much of the East was dry in 1987, and the return to more normal moisture levels in 1992 diminished a trend toward increased irrigation in the East.

Irrigated Crops

Most major crops are irrigated to some degree, but the number of acres and percentage of acres irrigated vary widely from crop to crop. Crops that have the greatest estimates for consumptive use and conveyance loss are not currently available. Thus the available estimates are rough approximations of actual conditions.

Water Withdrawals

Irrigation is by far the largest consumptive water user in the United States, particularly in the West. The quantity of water withdrawn for irrigation during 1990 was an estimated 137,000 million gallons per day, or 153 million acre-feet, which represents 40 percent of total U.S. freshwater use for all offstream categories. Irrigation withdrawals as well as acres irrigated during 1990 were about the same as during 1985. Water withdrawal and consumptive use information is summarized by water resource region and by state in Tables 3.2 and 3.3, respectively.

The nine western water resources regions, led by the Pacific Northwest region, accounted for 90 percent of the total water withdrawn for irrigation during 1990 (Table 3.2). In the eastern regions, most of the water withdrawn for irrigation was in the Lower Mississippi and South Atlantic–Gulf regions, which together had about 2,400 million gallons per day more water withdrawn during 1990 than during 1985.

Most states rely on a combination of surface and ground water supplies for irrigation purposes (see Table 3.3). Surface water accounted for 63 percent of total irrigation withdrawals in 1990. States with the highest share of surface water withdrawals include California, Montana, Wyoming, Oregon, Washington, and Utah.

Ground water is the primary supply source for irrigation in about half of the states (Table 3.3). Total ground water withdrawals were largest in California,

TABLE 3.2 Irrigation Water Use by Water Resources Region, 1990

Region	Irrigation System (thousand acres)		Thousand acre-feet per year: Withdrawals, by Sources, Fresh Water					
	Spray	Flood	Ground	Surface	Total	Reclaimed Wastewater	Conveyance Loss	Consumptive Use, Fresh Water
New England	49	12	9.9	124	134	0	0	134
Mid-Atlantic	347	3.6	114	106	221	0	2.8	188
South Atlantic–Gulf	2,660	1,140	2,580	2,420	4,990	264	76	3,570
Great Lakes	537	0.9	148	177	325	0	0	308
Ohio	166	0.5	31	45	76	0.3	0.6	67
Tennessee	31	0.1	4.2	26	30	0.4	0	21
Upper Mississippi	838	11	397	42	440	0.1	0.1	408
Lower Mississippi	977	3,800	6,990	1,290	8,280	0.8	672	6,160
Souris-Red-Rainy	98	22	63	47	110	0	1.2	98
Missouri Basin	4,880	7,950	8,070	19,700	27,800	3.4	10,100	12,300
Arkansas-White-Red	2,270	3,520	7,400	2,010	9,410	10	891	7,750
Texas-Gulf	1,710	2,680	4,450	1,270	5,720	34	383	4,820
Rio Grande	354	1,030	1,810	4,120	5,930	0.7	1,200	3,570
Upper Colorado	233	1,330	36	7,350	7,390	0.5	1,790	2,510
Lower Colorado	427	1,100	2,510	4,280	6,800	205	1,210	4,560
Great Basin	570	1,370	1,580	5,480	7,060	58	1,530	3,490
Pacific Northwest	4,210	3,280	8,800	26,800	35,600	12	10,800	13,100
California	2,310	7,300	11,900	19,800	31,700	143	1,960	21,700
Alaska	1.4	0	0.1	0.5	0.6	0	0.1	0.3
Hawaii	115	12	224	622	846	6.9	143	586
Caribbean	21	14	60	97	157	0	19	102
Total	22,800	34,600	57,200	95,900	153,000	740	30,800	85,400

Source: Solley et al., 1993.

TABLE 3.2 (continued)

	Million gallons per day				
	Withdrawals, by Source				
	Fresh Water				
Region	Ground	Surface	Total	Conveyance Losses	Consumptive Use, Fresh Water
New England	8.8	111	120	0	120
Mid-Atlantic	102	95	197	2.5	168
South Atlantic–Gulf	2,300	2,160	4,450	68	3,180
Great Lakes	132	158	290	0	274
Ohio	28	40	68	0.5	59
Tennessee	3.8	23	27	0	19
Upper Mississippi	354	38	392	0.1	364
Lower Mississippi	6,230	1,150	7,380	600	5,490
Souris-Red-Rainy	56	42	98	1.1	87
Missouri Basin	7,200	17,600	24,800	9,010	10,900
Arkansas-White-Red	6,600	1,790	8,390	794	6,910
Texas-Gulf	3,970	1,130	5,100	342	4,300
Rio Grande	1,620	3,670	5,290	1,070	3,180
Upper Colorado	32	6,560	6,590	1,600	2,240
Lower Colorado	2,240	3,820	6,060	1,080	4,070
Great Basin	1,410	4,890	6,300	1,360	3,110
Pacific Northwest	7,850	23,900	31,800	9,670	
California	10,600	17,700	28,300	1,750	19,400
Alaska	0.1	0.5	0.6	0.1	0.3
Hawaii	200	555	755	127	523
Caribbean	54	87	140	17	91
Total	51,000	85,500	137,000	27,500	76,200

Source: Solley et al., 1993.

TABLE 3.3 Irrigation Water Withdrawals and Consumptive Use, Major Irrigation States, 1990

	Withdrawals[a]				Consumptive Use[b]	
State	Irrigation Total (million acre-feet)	Surface Water–Bureau of Reclamation (percentage of irrigation water withdrawn)	Surface Water–Private Suppliers (percentage of irrigation water withdrawn)	Ground Water–All Suppliers (percentage of irrigation water withdrawn)	Irrigation Total (million acre-feet)	Irrigation's Share of Total Consumptive Use (percentage)
California	31.3	20	42	38	21.8	93
Texas	9.5	5	30	66	8.0	79
Idaho	20.9	44	21	35	6.8	99
Colorado	13.0	8	70	22	5.6	94
Kansas	4.7	2	3	95	4.5	92
Nebraska	6.8	13	15	71	4.4	93
Arkansas	5.9	0	18	82	4.4	94
Arizona	5.9	36	25	39	4.0	82
Oregon	7.7	25	67	8	3.4	95
Washington	6.8	70	17	12	2.9	92
Wyoming	8.0	18	79	3	2.9	95
Florida	4.2	0	48	52	2.8	79
Montana	10.1	11	88	1	2.2	93
Utah	4.0	9	77	14	2.2	87
New Mexico	3.4	21	33	46	2.0	86
Nevada	3.2	9	60	31	1.6	86
Mississippi	2.1	0	7	93	1.5	74
Louisiana	0.8	0	36	64	0.7	39
Georgia	0.5	0	40	60	0.5	54
Oklahoma	0.7	6	12	82	0.4	58
All other states	3.9	6	45	49	3.0	25
United States	153.0	20	43	37	85.4	81

[a] Withdrawal and consumptive use estimates are from the U.S. Geological Survey. They include freshwater irrigation on cropland, parks, golf, and other recreational lands.
[b] States are ranked on basis of total irrigation consumptive use.

Source: Solley et al., 1993.

Texas, and Idaho. Ground water as a share of irrigation withdrawals was highest in Kansas, Mississippi, Arkansas, Oklahoma, and Nebraska. Irrigated agriculture has contributed to declining aquifers in many areas.

In 1985, agriculture accounted for 42 percent of all freshwater withdrawals in the United States, or a total of 141 billion gallons per day, of which 97 percent was for irrigation and 3 percent was for livestock production. Freshwater withdrawals for agriculture are used mainly for crop production, with 98.4 percent of surface water and 93.8 percent of ground water used in irrigating cropland (Solley et al., 1988).

The trend in water used for all purposes for 5-year intervals from 1950 to 1990 is shown in Table 3.4. Included are withdrawals, source of water, reclaimed wastewater, consumptive use, and instream use (hydroelectric power). Table 3.4 also estimates the percentage increase or decrease in withdrawals between 1985 and 1990. After continual increases in the nation's water use from 1950 to 1980, offstream and instream uses were less during 1985 than during 1980. Total withdrawals were about 10 percent less during 1985 than during 1980, and the 2 percent increase from 1985 to 1990 is the result of increases in surface and ground water withdrawals of 1 and 9 percent, respectively. The fact that the 1990 withdrawal estimates are only slightly higher than the 1985 estimates tends to confirm the overall decline in water use from the peak of 1980.

The increase in estimated ground water withdrawals from 1985 to 1990 was partly the result of decreased availability of surface water. Surface water withdrawals for irrigation increased progressively for the years reported from 1960 to 1985 and decreased 6 percent from 1985 to 1990. It is expected that surface water withdrawals in the Pacific Coast and Pacific Northwest will remain at current levels or will decline as reallocations take place from agricultural use to streamflow maintenance to restore anadromous fish populations.

Water application varies from about 30 inches per year for crops such as rice and alfalfa to less than 10 inches per year for soybeans (Table 3.5). The amounts vary from region to region and from year to year depending on climatic conditions (especially temperature), precipitation, and irrigation practices. There is no direct annual measure of irrigation water applications, but 5 years of census and postcensus survey data suggest some trends (U.S. Department of Commerce, 1994). The east-west contrast in application rates has narrowed, with Atlantic states using almost twice as much water per acre in 1988 as in 1969. Despite increasing application rates in the East, national average application rates, as well as application rates for several major crops, have declined.

Consumptive Use

Consumptive use of fresh water in the United States totaled about 105 million acre-feet in 1990. Irrigation, the dominant consumptive water use, accounted for 85 million acre-feet, or 81 percent of the U.S. total. Consumptive use as a

TABLE 3.4 Trends of Estimated Water Use in the United States, 1950 to 1990

	Year				
	1950[a]	1955[a]	1960[b]	1965[b]	1970[c]
Population in millions	(150.7)	(164.0)	(179.3)	(193.8)	(205.9)
Offstream use					
Total withdrawals	180	240	270	310	370
Public supply	14	17	21	24	27
Rural domestic and livestock	3.6	3.6	3.6	4.0	4.5
Irrigation	89	110	110	120	130
Industrial					
Thermoelectric power use	40	72	100	130	170
Other industrial use	37	39	38	46	47
Source of water					
Ground					
Fresh	34	47	50	60	68
Saline	—[d]	0.6	0.4	0.5	1
Surface					
Fresh	140	180	190	210	250
Saline	10	18	31	43	53
Reclaimed wastewater	—[d]	0.2	0.6	0.7	0.5
Consumptive use	—[d]	—[d]	61	77	87[e]
Instream use					
Hydroelectric power	1,100	1,500	2,000	2,300	2,800

TABLE 3.4 (continued)

	1975[f]	1980[f]	1985[f]	1990[f]	Change from 1985 to 1990 (%)
Population in millions	(216.4)	(229.6)	(242.4)	(252.3)	4
Offstream use					
Total withdrawals	420	440[g]	399	408	2
Public supply	29	34	36.5	38.5	5
Rural domestic and livestock	4.9	5.6	7.79	7.89	1
Irrigation	140	150	137	137	0.3
Industrial					
Thermoelectric power use	200	210	187	195	4
Other industrial use	45	45	30.5	29.9	–2
Source of water					
Ground					
Fresh	82	83[e]	73.2	79.4	8
Saline	1	0.9	0.652	1.22	87
Surface					
Fresh	260	290	265	259	–2
Saline	69	71	59.6	68.2	14
Reclaimed wastewater	0.5	0.5	579	0.750	30
Consumptive use	96[e]	100[e]	92.3[e]	94.0[e]	2
Instream use					
Hydroelectric power	3,300	3,300	3,050	3,050	8

Note: Data for 1950 to 1980 adapted from MacKichan (1951, 1957), MacKichan and Kammerer (1961), Murray(1968), Murray and Reeves (1972, 1977), and Solley et al. (1983, 1988). The water use data are in thousands of million gallons per day and are rounded to two significant figures for 1950-1980, and to three significant figures for 1985 to 1990; percentage change is calculated from unrounded numbers.

[a] 48 states and District of Columbia.
[b] 50 states and District of Columbia.
[c] 50 states, District of Columbia and Puerto Rico.
[d] Data not available.
[e] Fresh water only.
[f] 50 states, District of Columbia, Puerto Rico, and Virgin Islands.
[g] Revised.

TABLE 3.5 Average Depth of Irrigation Water Applied per Season, 1969 to 1994, by Region and Crop (inches)

	1969[a]	1974[a]	1979[b]	1984[b]	1988[b]	1990[c]
Region						
Atlantic[d]	8.5	11.5	15.0	16.5	15.5	15.5
North Central[e]	7.5	8.0	9.5	9.5	10.5	9.0
Northern Plains	16.0	17.0	15.5	13.2	14.5	14.0
Delta States	15.5	17.5	26.0	17.5	18.0	16.5
Southern Plains	18.0	18.5	17.0	16.5	17.0	16.5
Mountain States	30.5	28.5	24.0	24.5	24.5	24.0
Pacific Coast	33.0	34.0	32.0	34.0	34.5	34.5
United States[f]	25.5	25.0	22.5	22.5	22.5	22.5
Crop						
Corn for grain	18.5	19.5	17.0	16.0	16.0	15.5
Wheat	23.0	24.0	20.5	16.5	16.0	15.5
Rice	28.0	28.5	33.5	33.5	32.5	31.5
Soybeans	12.0	11.5	14.0	9.5	10.0	8.5
Cotton	23.0	25.5	24.0	25.0	24.5	23.0
Alfalfa hay	32.5	30.5	26.0	28.0	29.0	28.5

TABLE 3.5 (continued)

	1991[c]	1992[c]	1993[c]	1994[g]
Region				
Atlantic[d]	16.0	16.0	16.5	16.5
North Central[e]	9.5	10.0	8.0	10.0
Northern Plains	14.0	13.5	11.5	14.5
Delta States	15.5	16.5	15.5	15.5
Southern Plains	15.0	16.0	16.0	16.0
Mountain States	24.0	24.0	23.0	24.0
Pacific Coast	34.5	34.5	33.0	35.0
United States[f]	21.5	21.5	20.0	21.5
Crop				
Corn for grain	15.0	15.0	13.5	15.5
Wheat	14.5	14.5	14.0	14.5
Rice	30.5	30.0	30.5	30.5
Soybeans	7.0	8.0	6.5	7.5
Cotton	21.0	23.0	20.0	21.5
Alfalfa hay	27.5	27.5	26.5	27.5

Note: Values are rounded to the nearest 0.5 inches.

[a] Census of Agriculture.

[b] Estimates constructed by state, by crop, from Farm and Ranch Irrigation Surveys (FRIS) (U.S. Department of Commerce, 1990, 1986, 1982a) and Economic Research Service estimates of irrigated area.

[c] Aggregated from Farm and Ranch Irrigation Surveys state/crop application rates adjusted to reflect annual changes in precipitation. Sensitivity to precipitation is estimated as a function of average precipitation and soil hydrologic group.

[d] Northeast, Appalachian, and Southeast production regions.

[e] Lake States and Corn Belt farm production regions.

[f] Includes Alaska and Hawaii.

[g] Forecast using precipitation records through May 1994.

Source: Economic Research Service, USDA.

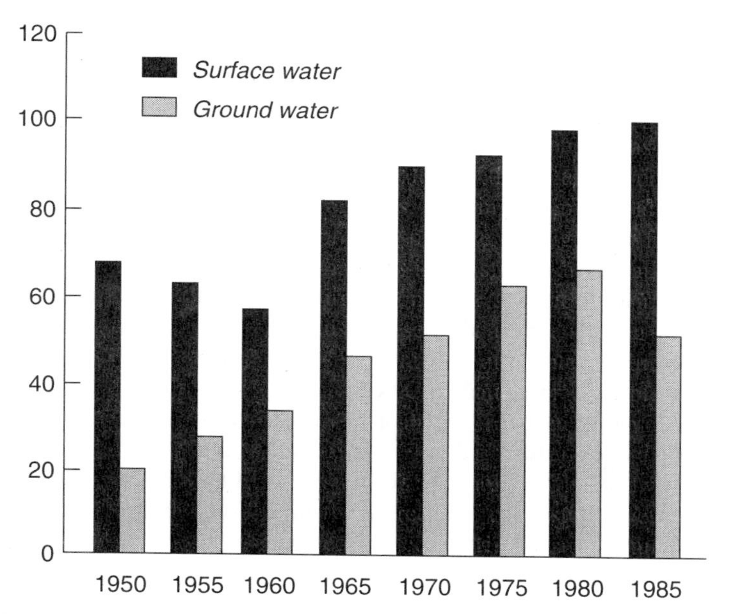

FIGURE 3.4 Trends in irrigation water use from surface and ground water. Source: Bajwa et al., 1992.

percentage of withdrawals was 56 percent for the irrigated sector, compared with 17 percent for public and rural supplies, 16 percent for industries other than thermoelectric, and just 3 percent for thermoelectric. Total consumptive water use for irrigation increased by about 60 percent between 1960 and 1980, reflecting the rapid expansion of irrigation in the West (Gollehon et al., 1994).

Irrigation consumptive use in the 20 major irrigation states accounted for 96 percent of the national total. California had the greatest irrigation consumptive use, followed by Texas, Idaho, and Colorado. Combined, these four states accounted for nearly half of the total irrigation consumptive use in the United States. Five of the top 20 major irrigation states—Arkansas, Florida, Mississippi, Louisiana, and Georgia—are in humid areas. Figure 3.4 highlights trends in use of surface and ground water for irrigation. Total water use in irrigated agriculture increased during the period from 1950 to 1980, but declined by 7 percent in 1985 despite continued growth in irrigated acreage nationwide. Reduced water use per irrigated acre reflects lower water applications in humid irrigated areas, a shift to less-water-intensive crops, and a reduction in irrigated cropland in some of the highest-water-using areas. Although surface water use increased slightly in 1985, declines in ground water use were greater than the increases in surface water use (Bajwa et al., 1992).

Irrigation Technology

Traditional irrigation technologies, such as furrow and border irrigation, rely on gravity to deliver water to crops and require substantial volumes of water over a short period of time. Irrigation using these traditional technologies is typically infrequent (once every 2 to 3 weeks, or even less). Modern irrigation technologies such as sprinkler, center pivot sprinkler, and microirrigation rely on energy and closed systems to deliver water to plants. These technologies allow more frequent and smaller irrigation input, improve irrigation distribution uniformity, and reduce water losses in deep percolation and runoff. In essence, the output produced with a given amount of water diverted is increased with these modern technologies.

Surface irrigation is still the most common form of irrigation in most states, particularly in the West, but sprinkler irrigation has been increasing rapidly since the 1950s and is used for field crops, fruits, and vegetables. The acreage of sprinkler irrigation increased by 9 percent from 1984 to 1988. Surface irrigation acreage remained almost level during the same period, which allowed sprinkler systems to increase from 37 percent of all systems in 1984 to 39 percent in 1988. The availability of aluminum and plastics was a significant factor in making sprinkler irrigation systems practical and economical.

Sprinkler technology has had a great impact on agricultural production in the West and the Midwest. In contrast, lands on the Mississippi Delta continue to be irrigated with traditional gravity-powered technology. For example, in Mississippi, where water is inexpensive, furrow irrigation is the principal system.

Drip irrigation, which is spreading very quickly in Florida and California, was introduced in the United States in the early 1970s and is currently in use on 1.5 million acres (Boggess et al., 1993). The technology tends to be adopted with high-value fruit and vegetable crops in locations with sandy soils, uneven terrain, and high water costs. Table 3.6 presents regional distributions of irrigated acreage by technology for five periods from 1960 to 1985.

There is a significant shift occurring in irrigation technology for crops and turf. The transition is to sprinkler and microirrigation. Surface irrigation is being modernized by laser leveling and the adoption of surface automation. Electronic controllers and sensors improve the control and management of irrigation systems. The key forces for adopting new technology are labor availability and cost, energy, limited water availability, and environmental concerns. The constraints to adoption are availability of capital, low-cost existing irrigation systems, user-unfriendly technology, limited management skills, and institutions that fail to provide incentives to conserve water.

Irrigation scheduling is an important technology for the effective use of limited resources. The monitoring of soil water with tensionmeter, gypsum blocks, neutron probles, and other sensors have been developed but slow to be adopted. Consultants are often providing scheduling services which include the

TABLE 3.6 Acreage Under Different Irrigation Technologies over Time for Selected States in the United States (thousands of acres)

	Gravity					Sprinkler						
	1960	1970	1975	1980	1986	1960	1970	1975	1980	1985	1980	1985
California	7,100	7,200	7,200	7,900	7,500	900	1,450	1,559	2,025	2,589	305	350
Hawaii	119	113	97	125	146	9	17	27	50	123	29	108
Florida	520	1,040	1,097	1,452	1,058	228	545	821	816	837	36	298
Georgia	2	0	0	0	0	102	145	230	1,013	1,080	5	45
Nebraska	2,415	3,536	3,973	4,509	4,404	250	866	1,642	3,128	3,793	0.1	0
Texas	5,152	6,660	6,700	5,620	4,617	610	1,700	1,918	2,197	2,141	2	0
Washington, Oregon	690	2,100	1,922	1,444	1,365	354	1,200	1,609	2,367	2,577	6	13
Southwest[a]	2,772	1,985	3,304	3,461	3,536	37	180	207	294	294	4	42
West[b]	8,472	12,076	11,535	10,517	9,913	392	1,600	3,076	3,733	4,030	1	2
Kansas, Oklahoma	920	1,971	2,306	2,802	2,539	100	560	953	1,754	1,677	0.7	3
Michigan	0.2	0.2	1	0	0	71	139	112	302	422	9	36
Iowa, Illinois	30	44	41	23	5	45	73	104	359	438	1	2
Northeast[c]	6	0.7	0.4	4	97	126	206	336	327	226	2	7
Total	28,197	36,693	38,177	37,956	35,181	3,223	8,680	12,594	18,365	20,225	419	906

[a] Arizona, Nevada, and New Mexico.
[b] Maine, New Hampshire, Vermont, Massachusetts, Rhode Island, Connecticut, New York, Pennsylvania, New Jersey, Maryland, and West Virginia.
[c] Idaho, Montana, North Dakota, Wyoming, Utah, and Colorado.

Source: Casterline et al., 1989.

use of soil probes for the real-time measurement of soil water. A constraint for adopting computer irrigation scheduling programs is the time required to input data. There is a need for user-friendly programs and cost-effective methods to automatically collect data and minimize the hand entering of data.

The development of new technology for improved irrigation systems had been accomplished by many entrepreneurial and relatively small specialized companies. Often, new systems and management technology is developed by partnerships between the irrigator, industry, consultants, entrepreneurs, and/or state and federal researchers. The federal government is encouraging cooperative research and development agreements for enhancing the technology transfer of new concepts and systems. Demonstration projects and cost-sharing programs for target problem areas provide for improvement of water use in critical areas.

ECONOMICS OF IRRIGATED AGRICULTURE

Irrigated yields exceed those for dryland farming by an average of 54 percent for corn grown for grain, 97 percent for wheat, 33 percent for soybeans, and 67 percent for cotton (U.S. Department of Agriculture, 1986) (Table 3.7). Irrigated farms tend to be more highly capitalized than nonirrigated farms. They produce significantly more crop and livestock value per farm and have higher expenditures for agricultural chemicals, energy, and labor. The average irrigated farm has over twice as much invested in land and buildings and twice the value of machinery and equipment as nonirrigated farms. Thus irrigated agriculture is more directly affected by a changing economic and financial environment. As water costs rise, it becomes necessary to economize on water use and to select those agricultural enterprises that can be profitable with higher cost water (McNeely and Lacewell, 1978). The value of water to agriculture is dependent on the crops produced, crop response to water, crop prices, energy costs, soil productivity, and other production costs.

Irrigation Water Prices and Costs

The prices of most agricultural inputs are established in markets, where prices indicate relative scarcity through supply and demand. In contrast, irrigation water prices are typically not set in a market. Water prices usually reflect only the cost of supplying water and generally do not convey market signals. Irrigation water costs vary widely, reflecting different combinations of water sources, suppliers, distribution systems, and other factors (Gollehon et al., 1994).

The costs of providing on-farm surface water are relatively low. On-farm surface water pumps generally lift water less than 20 feet, resulting in low energy costs of $2 to $15 per acre-foot. Initial expenditures for surface water pumps can vary greatly depending on farm-specific conditions, but most systems cost $3,000 to $10,000 (Gollehon et al., 1994).

TABLE 3.7 Yields on Irrigated Lands as a Percentage of Nonirrigated Yields, 1984

	U.S. Average	20 Principal Irrigated States[a]
Corn for grain	154	201
Corn for silage	162	150
Sorghum	127	182
Wheat	197	197
Barley	203	208
Soybeans	133	138
Other beans	133	131
Rice	—	—
Alfalfa	163	176
Other hay	119	127
Cotton	167	187
Sugar beets	121	—
Tobacco	120	—
Potatoes	150	195

[a] Includes the 17 western states plus Arkansas, Florida, and Louisiana.

Source: Calculated from data in U.S. Department of Agriculture, 1986, p. 21.

TABLE 3.8 Labor Requirements and Capital Costs for the Various Irrigation Methods[a]

System	Labor requirement (h/acre-irrigation)	Capital Costs[b] ($/acre)
Surface		
Border	0.2-1.0	120-400
Furrow	0.4-1.2	160-500
Corrugation	0.4-1.2	100-200
Level basin	0.1-0.5	200-500
Sprinkler		
Fixed		
Solid set portable	0.2-0.5	400-1200
Permanent	0.05-0.1	400-1200
Periodic move		
Hand move	0.5-1.5	100-300
End tow	0.2-0.5	180-350
Side roll	0.2-0.7	180-350
Moving		
Traveler	0.2-0.7	200-400
Center pivot	0.05-0.15	200-400
Linear move	0.05-0.15	300-500
Micro		
Drip	0.15	250-1000
Subsurface	0.15	250-1000
Bubbler	0.15	250-1000
Spray	0.15	250-1000

[a] Modified from Turner and Anderson (1980) and Lord et al. (1981).
[b] Excluding cost of water supply, pump, or power unit.

Source: Council for Agricultural Science and Technology, 1988.

Production costs associated with ground water pumping are generally higher and reflect both the variable cost of extraction and the fixed cost of access. Total energy expenses for irrigation pumping reached $1.05 billion in 1988, up 5 percent from 1984. Average expenditures per acre were slightly lower in 1988 than in 1984, reflecting shifts to more efficient application systems and changes in the mix of irrigated crops. Of the five types of energy used for pumping irrigation water—electricity, natural gas, liquefied petroleum (LP) gas, diesel, and gasoline—electricity (56 percent), diesel (21 percent), and natural gas (17 percent) dominated in 1988. Electricity and natural gas declined in importance, while the use of diesel grew by 4 percent between 1984 and 1988. Average energy expenditures by state range from $11 to $105 per acre (Gollehon et al., 1994).

The major considerations in selecting an irrigation system involve capital and operating costs, crop(s) to be irrigated, and expected crop yield and quality. Ranges of installed capital costs for the various types of irrigation systems are given in Table 3.8. The increase in crop returns over the useful life of a system must be great enough to repay the capital and annual operating costs. Labor and energy are the two major components of operating costs.

Labor requirements for irrigation systems vary greatly. Automated systems, such as automated microirrigation and center pivot systems, have relatively low labor requirements. Labor requirements for the main irrigation methods tabulated by Turner and Anderson (1980) and Lord et al. (1981) are given in Table 3.8. Annualized costs are not shown because of the wide range in the expected life of the various systems or system components. The data do show the large differences in capital costs encountered because of differences in water sources, field shapes and topography, soils, and large differences in labor requirements because of automation.

More than 60 percent of the West's irrigated lands use gravity to distribute water. Sprinkler irrigation systems, used on about 36 percent of the West's irrigated lands, tend to be considerably more expensive than gravity systems. Center pivot systems, for instance, cost more than $300 per acre installed (three times more than gravity systems) and about $15 for energy for each acre-foot applied. On the other hand, center pivots require very little labor to operate. Mobile trickle systems that attach trail lines with emitters to center pivots or other mobile sprinkler systems have been introduced into the High Plains to reduce evaporation losses and energy use (Frederick and Hanson, 1982, pp. 158-165). In the High Plains, a large number of LEPA (low-energy precision application) systems were also installed in place of higher-pressure sprinkler systems (Bryant and Lacewell, 1988).

Economic factors, especially crop and energy price levels, will be important to the future growth of irrigated agriculture. Because their yields and production costs are generally higher, irrigators' profits are more sensitive than those of other farmers to agricultural prices. High crop price levels encourage yield-increasing investments, of which irrigation is an important option. On the other hand, high energy prices are likely to affect irrigators more adversely than other farmers.

Rising real crop price levels can offset higher water costs and encourage additional ground water pumping. But the lure of additional profits from irrigation would not alter the trend for more of the region's water to flow to municipalities and industries where water values are much higher than in irrigation. Nor would it do much to increase the flow of public funds for irrigation projects that are criticized for their adverse impacts on the environment and instream water uses as well as for their questionable economics. Higher crop prices, however, would make irrigators more inclined and better able to respond to rising water costs with investments designed to increase the return to water (Frederick, 1988a).

Value of Irrigation Water and Water Marketing

The value of water in agriculture (principally for irrigation) is often estimated by calculating the value of the last unit applied. The value of irrigation water provides a base of comparison to the value of water in other uses. The Office of Technology Assessment (1983) indicated that the value of an acre-foot of water used in irrigation ranges from $9 to $103 (Boggess et al., 1993). Typically, horticultural crops and other high-value crops are associated with the highest value of water, while pasture and alfalfa are associated with the lowest values.

The value of water in agriculture is generally less than in industrial and municipal uses, and the price elasticity of demand for industrial and municipal water is more inelastic than that for agriculture. This means that when the need for additional supply arises for municipal and industrial users, they can offer higher prices for water than can agriculture (Boggess et al., 1993). The difference between the value of water in agricultural, industrial, and municipal uses is largely due to the limited use of markets for allocation of water among users. For example, water from reservoirs and transport facilities built by the Bureau of Reclamation is allocated according to water rights based on past use rather than willingness to pay, and trading has been constrained.

Because it is so expensive to develop additional water supplies, only the higher-value water uses are likely to be justified economically. The implied average cost of adding an acre-foot to annual water supplies through recommended conservation measures is between $1,000 and $2,500. While this cost is not out of line with prices paid for water rights in some areas of the West, it is well above the value of water for most agricultural uses. About 90 percent of the consumptive use of western irrigation water is applied to crops for which the marginal value of water is less than $100 an acre-foot; nearly one-half is for crops with marginal water values of $30 an acre-foot or less (Gibbons, 1986). Thus the present value of a permanent increase in net water supplies is much less than $1,000 per acre-foot for all but the higher-value crops. The trend in the future thus will be toward higher-valued crops—for instance, orchard crops and vegetables rather than hay.

Institutional constraints also limit the flow of water to higher and more economically valuable uses. A market-oriented allocation system would allow those with higher-valued uses to bid water away from many lower-valued uses. Such an approach, however, often requires modification in water laws and institutions. For example, in some states such as California and Texas, farmers attracted by offers of high prices are selling water to municipal and industrial users. The higher water values often characteristic of nonagricultural uses have led to predictions that such transfers would lead to the demise of irrigation in large areas. However, such forecasts may not be accurate. Because of the current dominance of irrigation in western water use, large percentage increases in nonagricultural water uses can be met with relatively small percentage reductions in irrigation use. For instance, nearly 90 percent of the consumptive use of western water is for irrigation. Thus a 10 percent reduction in irrigation use would be sufficient to almost double the water available for municipal and industrial uses (Frederick, 1988b).

IRRIGATION AND THE ENVIRONMENT

The principal environmental issues relevant to irrigation are those concerned with the protection and management of water supplies and water quality. In the last 25 years, the public has become increasingly conscious of and concerned about environmental quality, endangered species, and public health and safety, and of the impacts of agricultural irrigation on these resources. Urban and suburban expansion into rural, agricultural regions has also given rise to conflicts over land use, waste disposal, recreational access, chemical use, and other issues. Environmental issues related to water consumption and water quality in landscape irrigation have become more prominent with the expansion of golf courses and the turfgrass sector generally. The relative significance of environmental issues associated with irrigation varies between regions of the country, but the types of environmental issues confronting irrigation generally are the same coast to coast.

Irrigation has been insulated in some ways from direct environmental regulation over the past 25 years. Most of the environmental laws and policies adopted in the period from the late 1960s to the early 1980s had little to do specifically with irrigation; rather they evolved in response to concerns over endangered species, wilderness preservation, point sources of contamination, and threats to ambient air and water quality. Some irrigation impacts were exempted explicitly from regulation, such as the exemption of irrigation return flow as an unregulated nonpoint source of pollution under the Clean Water Act. At the same time, laws and institutions pertaining specifically to irrigation had little to say about environmental issues. Instead, they were designed to advance other social goals such as the settlement of the West, reliability and affordability of irrigation water supplies, stabilization of the agricultural economy through crop payments, assistance to individual farmers affected by natural disasters such as drought or pests, and soil conservation.

However, the environmental impacts of activities associated with irrigated agriculture have been profound. Irrigation has contributed directly to losses of aquatic habitats and the decline of species that depend on them (Wilcove and Bean, 1994).[2] Runoff from irrigation is a significant source of water pollution in rivers, lakes, and estuaries (National Research Council, 1989).

The potential for conflict between irrigation and environmental goals is inherent in the fact that water is the limiting resource in irrigation and in aquatic ecosystems. Although irrigation has been largely exempt from the "command and control" environmental regulations applied to other industries, the trend in environmental policy is one of greater focus on, and control of, irrigation activities that have the potential to affect endangered species and their habitat, sensitive ecosystems such as wetlands, water quality, and public health. Irrigation's influence on the environment also is receiving public and political scrutiny with the growing concern over the costs of environmental protection and subsidies to natural-resource-based industries. Under the Bureau of Reclamation, repayment requirements for irrigators have been generous, with federal irrigation subsidies averaging in excess of 86 percent of construction costs (Wahl, 1989). In some cases, federal (or state) subsidized water is used to irrigate lands that in turn grow crops subsidized under federal commodity price support programs. In an era of increased competition for water supplies, and with state and the federal governments struggling with budget constraints, the costs of these programs is being called into question. The "externality" costs of pollution from agriculture are increasingly controversial, particularly where irrigators receive subsidized water.

One measure of growing public concern over the environmental impacts of irrigation is the number of laws and regulations that pertain to irrigation. Federal and state responses to environmental concerns about agriculture, especially irrigation, include efforts to control salinization and agricultural nonpoint sources of water pollution, water policies designed to protect instream flows and wetlands, and restrictions on the types and application of agricultural pesticides. Table 3.9 lists the federal programs related to water quality and agriculture. Most of these programs are obviously relevant to irrigated agriculture as well as to agriculture generally. The table does not include federal programs for species and habitat protection or other environmental issues relevant to agriculture. The committee cannot predict how the legislative pendulum will shift given the current, more conservative bent of Congress and emphasis on cost cutting, but its feeling is that while there will be some changes, the American people will not support a wholesale retreat from environmental protection. Key environmental issues with direct association to irrigated agriculture are instream flows and wetlands, salinity and drainage, water quality, and anthropogenically induced climate change.

TABLE 3.9 Major Federal Programs Related to Water Quality and Agriculture

U.S. Department of Agriculture

1985 Food Security Act Provisions

Conservation Reserve Program (CRP) provides annual rental payments to land owners and operators who voluntarily retire highly erodible and other environmentally critical lands from production for 10 years. It also provides technical assistance and cost-sharing payments of up to 50 percent of the cost of establishing a soil-conserving cover on retired land. Over 30 million acres of cropland have been enrolled.

Sodbuster provisions require that farmers who convert highly erodible land to agricultural commodity production do so under an approved conservation system, or forfeit eligibility for USDA program benefits.

Swampbuster provisions bar farmers who convert wetlands to agricultural commodity production from eligibility for USDA program benefits, unless USDA determines that conversion would have only a minimal effect on wetland hydrology and biology.

Continuing Assistance Programs

Agricultural Conservation Program (ACP) provides financial assistance through the Agricultural Stabilization and Conservation Service (ASCS) to farmers for implementing approved soil and water conservation and pollution abatement practices. Except for Water Quality Special Projects, conservation priorities are set by states and counties based on local soil and water quality problems. The program was initiated in 1936. ASCS also administers the Integrated Crop Management (ICM) program, a pilot ACP project to improve agrichemical management through cost-share assistance from crop advisory and soil testing services. The program was initiated in 1990.

Conservation Technical Assistance (CTA) provides technical assistance by the Soil Conservation Service (SCS) through Conservation Districts to farmers for planning and implementing soil and water conservation and water quality improvement practices. The program was initiated in 1936.

Small Watershed Program provides federal technical and financial help to local organizations for flood prevention, watershed protection, and water management. The program was initiated in 1954.

Resource Conservation and Development Program assists multicounty areas to enhance conservation, water quality, wildlife habitat, recreation, and rural development. The program was initiated in 1962.

Rural Clean Water Program is an experimental program implemented in 21 selected projects. It provides cost sharing and technical assistance to farmers voluntarily implementing best management practices to improve water quality. The program was initiated in 1980; it ends in 1995.

Water Bank Program provides annual payments for preserving wetlands in important migratory waterfowl nesting, breeding, or feeding areas. The program was initiated in 1970.

Environmental Protection Agency

FIFRA Pesticide Programs

The Federal Insecticide, Fungicide, and Rodenticide Act (FIFRA) gives EPA responsibilities for registering new pesticides and for reviewing and re-registering existing pesticides to ensure that, when used according to label directions, they will not present unreasonable risks to human health or the environment. Under FIFRA provisions, EPA may restrict or cancel use of any pesticide determined to be a potential hazard to human health or the environment.

National Survey of Pesticides in Drinking Water Wells

The National Survey tested for the presence and concentration of 127 commonly used agricultural chemicals in 1,350 statistically selected wells in all states. Water samples were analyzed and questionnaires filled out by well owners, operators, and local area experts on well construction locale and cropping and pesticide use patterns.

Safe Drinking Water Act Programs

The Safe Drinking Water Act (SDWA) requires EPA to publish maximum contaminant levels (MCLs) for any contaminants, including pesticides, that may have adverse health effects in public water systems (those serving over 25 persons or with 15 connections). Standards established by EPA under the SDWA are also being used as guidelines to assess contamination of ground water in private wells. The EPA also sets nonregulatory health advisory levels on contaminants for which MCLs have not been established.

Clean Water Act

1987 Water Quality Nonpoint Programs. Section 319 of the Clean Water Act requires states and territories to file assessment reports with EPA identifying navigable waters where water quality standards cannot be attained or maintained without reducing nonpoint-source pollution. States must also file management programs with EPA identifying steps that must be taken to reduce nonpoint pollution in those waters identified in the state assessment reports. The act authorizes up to $400 million total in federal funding for implementing the programs. To date, 43 states and territories have submitted nonpoint-source pollution assessments to EPA, and 36 have submitted final management programs.

Clean Lakes Programs. Section 314 of the act requires states to submit assessment reports on the status and trends of lake water quality, including the nature and extent of pollution loading from point and nonpoint sources. Also, methods to control pollution and to protect/restore the quality of lakes impaired or threatened by pollution must be described.

National Estuary Program. Section 320 of the act provides for identification of nationally significant estuaries threatened by pollution, preparation of conservation and management plans, and federal grants to prepare the plans. Twelve major estuaries have planning underway.

Near Coastal Waters Strategy

Through its Near Coastal Waters Strategy, EPA is integrating its water quality programs to target priority programs and prevent pollution in near coastal waters. This includes the implementation of nonpoint-source management programs in coastal counties and will, in several cases, encompass accelerated implementation of agricultural conservation programs.

Regional Water Quality Programs

The EPA and other federal agencies are cooperating on several regional programs to reduce nonpoint-source pollution, including the Chesapeake Bay Program, the Colorado River Salinity Control Program, the Great Lakes Program, the Gulf of Mexico Program, and the Land and Water 201 Program in the Tennessee Valley Region.

U.S. Geological Survey

National Water Quality Assessment Program

Since 1986 the NAWQA program has conducted assessments of national and regional status of ground water resources and monitors trends in factors that can affect ground water quality. Agrichemical nonpoint-source contamination problems are under study in seven pilot projects.

Regional Aquifer Systems Analysis Program

The RASA program was established in 1978 to gather data on the quantity of water resources available in the nation's aquifers. RASA's objectives for each aquifer system study are to determine the availability and chemical quality of stored water and discharge-recharge characteristics, and to develop computer simulation models that may assist in understanding the ground water flow regime and changes brought about by human activities. Twenty-eight aquifer systems have been identified for study, fourteen of which have been completed.

Federal-State Cooperative Program

USGS supports local efforts to collect data on ground and surface waters through cost-sharing arrangements with state and local governments. For example, USGS has provided support for mapping state aquifers and for monitoring pesticide contamination problems and has assisted in developing wellhead protection programs.

Source: Carlson et al., 1993.

Instream Flows and Wetlands

Problems related to instream flows and wetland ecosystems occur in every region of the country where significant quantities of surface and ground water are withdrawn for irrigation. Dams and diversions for surface supplies reduce instream flows, altering the natural hydrograph and affecting water temperature and flow regimes, trapping sediments, and changing water quality. In addition to obstructing the passage of migratory fish, these changes degrade spawning and rearing habitats in the stream and riparian areas. The draining and filling of wetlands for irrigation have significant impacts on waterfowl and other aquatic species that use these habitats for nesting and breeding and also increase the potential for sedimentation and water pollution.

In California, for example, construction of the Friant Unit of the Central Valley Project resulted in the dewatering of the San Joaquin River for a 50-mile reach below Friant Dam. As a result of dams and diversions for irrigation, water supplies available for fish and wildlife habitat have been greatly reduced. Ninety-two percent of the historic wetland acreage in the San Joaquin Valley has been converted to irrigated agriculture. (San Joaquin Valley Drainage Program, 1990). In Idaho, ground water pumping by irrigators along the Big Lost River over the last 15 years has caused the dewatering of the lower reach of the river and lowered ground water levels precipitously (High Country News, 1995). Large-scale irrigation projects constructed by the Bureau of Reclamation have drastically altered habitat conditions in major river basins across the West, including the Platte River, the Colorado River, the Columbia River, and the Snake River.[3] Many fish and other aquatic species that depend on habitat values in these rivers are listed as threatened or endangered under federal and/or state endangered species laws, although it must be noted that irrigation withdrawals are only one factor among many (e.g., hydroelectric power generation) that contribute to instream flow problems.

Salinity and Drainage

Salinity and drainage problems arise from natural hydrological and geochemical factors—the earth's rocks and soils contain mineral salts, which are released via normal chemical weathering processes. Irrigation in areas rich in such salts can concentrate the salts in water and soils (surface evaporation and transpiration by plants both act to move water into the atmosphere, leaving concentrated salts behind). The major dissolved mineral salts at issue are sodium, calcium, magnesium, potassium, chorine, SO_4, HCO_3, CO_3, and NO_3. Over time, salts concentrated in soils can hinder plant germination, seeding, and growth and undermine the yield and quality of plants. Saline drainage water can have adverse effects on water quality and, in turn, harm wildlife populations and make the water less desirable for other users.

About 30 percent of the land in the conterminous United States, much of it concentrated in the West, has a moderate to severe potential for salinity problems (Tanji, 1990). The upper Colorado River basin, the northern Great Plains, and California's San Joaquin Valley are examples of areas that suffer salinity and drainage problems. The accumulation of salts in soils depends on the salinity of the applied waters, the salinity of the native soil, and the rate at which salts are leached out of the root zone. A related problem is waterlogging of the soil: waterlogging in the root zone depends on the drainage characteristics of the soil, whether there is a restricting layer in the soil, and the soil's capacity for deep percolation. In poor conditions, waterlogging can occur relatively rapidly. In good conditions, irrigation may be practiced for decades, and even centuries, before surface drainage problems arise. Irrigation-induced salinization can be avoided by providing adequate drainage, but drainage is expensive and exacts an environmental price as well—it degrades water quality along its disposal route and in closed basins can render the terminus biologically uninhabitable (van Schilfgaarde, 1990).

Water Quality

Surface return flows and drainage from irrigation are a leading source of water pollution in rivers, lakes, streams, and estuaries nationwide. According to recent estimates, irrigated cropland in the West accounts for 89 percent of quality-impaired river mileage. Irrigated agriculture accounts for more than 40 percent of the pollution in lakes with impaired water quality (U.S. Environmental Protection Agency, 1992). In the arid West, low river flows can exacerbate pollution problems from irrigation because surface runoff and drainage often provide a significant portion of these flows. Pollutants mobilized and transported by irrigation return flows and drainage include naturally occurring trace elements (e.g. selenium, boron, molybdenum), nitrogen, and salts, as well as pesticides, herbicides, and other chemicals (U.S. Fish and Wildlife Service, 1992). Significantly, irrigation return flows are the most common source of pollution in national wildlife refuges (U.S. Environmental Protection Agency, 1992). While fewer data are available on the effects of agricultural drainage on species other than waterfowl, agricultural runoff is believed to affect adversely fish populations in many river reaches in the country (U.S. Fish and Wildlife Service, 1992). The trend toward a greater public policy focus on irrigation's impact on the environment is borne out by changes in various policies and institutions serving both irrigation and environmental goals. The mission of the Soil Conservation Service, now called the Natural Resources Conservation Service, has been modified and expanded over the past 50 years. It has gone from helping farmers prevent soil erosion to conducting activities and providing technical and financial support to farmers to conserve highly erodible and environmentally sensitive lands and protect water quality.[4] In 1987 the Bureau of Reclamation, historically the supplier of one-fifth of all irrigation (agricultural) water in the United States and manager of 45 percent of the West's water, announced the end of its mission of helping to settle the West through the construction and operation of major dams and diversions, and the beginning of a mission focused on resource management (Bureau of Reclamation, 1987). In 1992, the Central Valley Project Improvement Act (P.L. 102-575, Title XXXIV, 106 Stat. 4706) set aside 800,000 acre-feet of water previously delivered by the federal Central Valley Project to agricultural users for fish and wildlife habitat.[5] In addition, water users were required to pay surcharges on irrigation water to be used to finance environmental restoration.

In 1987, amendments to the Clean Water Act required states to assess the extent of nonpoint-source water quality impairment and to develop programs to manage nonpoint-source pollution. Section 319 of the act authorized $400 million in grants to states to assist in this effort (33 U.S.C. Section 1329). In addition, nonpoint-sources must be factored into the calculations that allocated pollution reduction responsibilities among dischargers for each water body that does not meet water quality standards (section 303; U.S.C. Section 1313). In 1990, amendments to the Coastal Zone Management Act, administered by the National Oceanic and Atmospheric Administration and the Environmental Protection Agency, required states with coastal zone management programs to develop programs for the control of nonpoint sources, including agriculture (Coastal Zone Act Reauthorization Amendments, 1990).

Several states have adopted programs or passed legislation to protect aquatic habitats and the species that depend on them. Minimum instream flow requirements, appropriations for instream rights, water transfer options, conservation easements, and other mechanisms are being employed to address problems concerning the quantity and quality of water available to fish and wildlife resources.

In the turfgrass sector, soil erosion and runoff during construction and the potential for leaching and runoff of nutrients and pesticides from established sites can lead to impacts on fish and wildlife habitats and aquatic systems. These impacts likely will continue to fall under the urban stormwater provisions of the Clean Water Act and sometimes state legislation.

Climate Change

The Second Scientific Assessment of Climate Change by the Intergovernmental Panel on Climate Change (IPCC) (1996) concludes for the first time that a global warming attributable to human activities is now evident in the historic record. Under a mid-range emission scenario, global mean surface temperature relative to 1990 is expected to increase by about 2°C by 2100, when the effects of greenhouse gas emissions and sulfate are considered.

Although beyond the time horizon that is the focus of this study, if it occurs, greenhouse warming is certain to have a major impact on water supplies. A warmer climate would accelerate the hydrologic cycle, increasing both the rates of precipitation and evapotranspiration. The regional impacts, however, are highly uncertain. Regional precipitation patterns, evapotranspiration rates, the timing and magnitude of runoff, and the frequency and intensity of storms would be affected. But the magnitude and sometimes even the direction of the changes for particular river basins and watersheds are uncertain. The range of likely changes in average annual precipitation associated with an equivalent doubling of atmospheric carbon dioxide for any given region might be on the order of plus or minus 50 percent (Schneider et al., 1990).

The hydrologic uncertainties are compounded because relatively small changes in precipitation and temperature can have sizable effects on the volume and timing of runoff, especially in arid and semiarid areas. For example, Nash and Gleick (1993) have speculated on the estimated impacts of alternative temperature and precipitation changes on annual runoff in several semiarid areas. In their scenario, with no change in precipitation, estimated runoff in these study areas declines by 3 to 12 percent with a 2°C increase in temperature and by 7 to 21 percent with a 4°C increase in temperature. A 10 percent increase in precipitation does not fully offset the negative impacts on runoff attributable to a 4°C increase in temperature in three of the five basins for which this climate scenario was studied.

The CO_2 fertilization effect will affect plant growth and possibly water supplies. Research results suggest that the increasing levels of atmospheric carbon dioxide (CO_2) levels will increase the growth and yield of C_3 plants (small grains, legumes, root crops, and most trees) by 34(+/-6) percent and C_4 plants (e.g., maize and sorghum) by 14 (+/-11) percent (Rosenberg et al., 1990). The impacts of the CO_2 fertilization effect on water supplies is less certain because of two counteracting effects. On the one hand, an increase in leaf and root areas has the potential to increase transpiration and, thereby, reduce runoff. A simulation analysis suggests that a 15 percent increase in the leaf area index (other things being constant) would increase summertime evapotranspiration from a wheat field in Nebraska by 5 percent. On the other hand, a rise in atmospheric CO_2 levels would increase stomatal resistance, the primary plant factor controlling evapotranspiration. Transpiration from a given leaf area declines as the stomatal resistance rises. In another simulation of the impacts of climate variables on the Nebraska wheat field, a 40 percent increase in stomatal resistance (other things being equal) reduces summertime evapotranspiration by 12 percent (Rosenberg et al., 1990).

In summary, the prospect of a global greenhouse warming introduces major new uncertainties and challenges for irrigators as well as for other farmers and water users. The allocation of water supplies among competing uses in response to any climate-induced shifts in hydrology and the response of irrigators to these changes is likely to be an important determinant of the future of irrigation.

THE TURFGRASS SECTOR

When precipitation is insufficient, turfgrass must be irrigated to provide the desired turf appearance and recuperative ability. Problems arise when there is an extended period of lack of precipitation or lack of availability of either ground water or surface water to allow for turf irrigation. The importance of turfgrass irrigation was most clearly realized during the drought period of 1976 to 1978 in the western United States, when extensive damage occurred.

Although turf was commercially recognized before World War II, the rapid growth and development of the turf industry occurred after the war. In 1965 turfgrass was a $4.3 billion industry (Turfgrass Times, 1965). By 1992 it had grown to a nearly $30 billion industry. The fixed asset value of turf is, of course, many times that annual expenditure. California, Florida, Michigan, New York, North Carolina, Pennsylvania, and South Carolina all have billion-dollar turf industries, and Illinois and Texas are very near this level. A survey of 2,309 golf courses in late 1984 by the Golf Course Superintendents Association of America (GCSAA) and the National Golf Foundation (NGF) provided statistical data on the acreage and cost of maintaining America's golf courses. Projecting the financial data obtained from the sample, it is estimated that $1.7 billion is spent each year for golf course maintenance and that the nation's courses had a maintenance equipment inventory valued at more than $1.8 billion (Prusa and Beditz, 1985). Even though the technology of turfgrass management has undergone tremendous development, it is still labor intensive. It is estimated that 380,000 people make their living directly from the care and maintenance of turf in the United States.

There are over 50 million home lawns and more than 14,000 golf courses in this country (Schroeder and Sprague, 1994). Water use rates for turfgrass vary widely, from 0.1 inch per day for foggy coastal climates to 0.45 inch per day for dry desert areas (Beard, 1982). A golf course may require a water source capable of supplying as much as 1.5 to 3.5 million gallons of water per week during the golf season (Jones and Rando, 1974).

Surface waters of all types are common direct sources of water for golf courses and other larger turfgrass areas. Frequently, small streams and major drainage channels may be dammed, excavated, or both, and the impounded water used to irrigate the golf course. Small reservoirs (less than 50 acre-feet in size) provide only 2 percent of the nation's total storage capacity. However, they are a significant source of water for golf courses and park areas. Water harvesting and storage in small ponds and reservoirs are increasingly becoming a major element in golf course design.

Treated effluent water, although not technically "surface water," is an alternative source of supplemental irrigation water. Because of its nutrient content, it is a particularly valuable source of irrigation water for sod farms and golf courses. The quality of the effluent depends on the source; therefore, it varies widely.

Advantages and disadvantages are associated with use of effluent (Watson, 1978). Effluent water or wastewater is used to irrigate several golf courses, including the Eisenhower Course at the Air Force Academy; Innisbrook at Tarpon Springs, Florida; Randolph Park at Tucson, Arizona; and some military courses. Development of multiple-plumbing systems to accommodate regular and effluent water for turf facilities is inevitable. Many golf courses already use such systems, and use of effluent water for golf course irrigation is mandatory in California where it is available (Thomas, 1994).

The availability of sufficient water of adequate quality and price in the future will pose a challenge to the turfgrass industry. In humid and subhumid areas, watering of home lawns often is restricted because municipal distributive systems have not kept up with the rapid expansion of suburban areas. For the same reasons, watering of turfgrass areas may be restricted in semiarid and arid regions. Alternative water sources that may be useful to the turfgrass industries include wastewaters, including treated sewage effluents; capture and impoundment of runoff waters; and dual water systems for turf facilities, including home lawns, to accommodate potable and nonpotable waters (Watson, 1985).

THE SPECIAL CASE OF INDIAN IRRIGATION

The legal, historical, and political framework for Indian irrigation and natural resource use is rooted firmly in the history and development of the United States. American Indians have a unique relationship with the United States which stems from the Constitution, Treaties, Executive Orders, Court decisions, and legislation enacted since the late eighteenth century through the present. The body of law created by these mechanisms establishes a framework for the implementation of the U.S. trust responsibility for the protection of Indian natural resources. As the twenty-first century approaches, the increased implementation of these treaty rights through the development of water for agricultural or nonagricultural enterprises is central to Indian economic development activity.

While much of the treaty making was completed more than a century ago, it is only now that many of the provisions of the treaties are coming to fruition. The securing of water supplies and other natural resources has implications for irrigated agriculture in the United States, particularly in the Northwest, Southwest and Missouri River basin. Today, American Indians own 2.7 million acres of cropland, of which 64 percent is irrigated. The total estimated income from Indian irrigation, both in private systems and BIA-administered programs exceeds $1 billion annually.[6]

The development of irrigated agriculture on Indian reservations and the forced transformation of Indian culture in the mid to late nineteenth century formed the core development vision of U.S. policy regarding American Indians. Reservations were set aside as homelands, whose purpose was envisioned as agricultural. Most tribal irrigation projects were authorized congressionally. Under the general appropriations for irrigation authorized by Congress, the irrigation systems that were built for tribes were refinements of earlier irrigation systems constructed by the tribes themselves prior to any assistance from the federal government (Bureau of Indian Affairs, 1975). In several instances, existing non-Indian projects were extended to meet the needs of the Indians. While all of this work was designed to "fulfill treaty stipulations with various Tribes," many irrigation systems on Indian reservations were constructed, improved, or extended by the federal government without consideration as to economic feasibility and repayment capability, a fact that is common to all irrigation projects constructed with federal funds during this century. In many cases, such projects were constructed without the consent of the Indians involved.

Table 3.10 presents a partial listing of the 71 statutorially authorized Indian irrigation projects. The Pick-Sloan program of the 1944 Flood Control Act also authorized the construction of Indian irrigation projects. To date, few Indian irrigation projects have been constructed under the Pick-Sloan program. It is significant to note that initial appropriations authorized were in most cases not sufficient to finish the project, nor to design the project for effective water delivery. In addition, funding did not cover routine operation, maintenance, and replacement activities. Nearly all of these projects have serious replacement, operations, and maintenance costs and other problems that have inhibited full agricultural development and effective water delivery.

There are also statutorily authorized power projects in conjunction with irrigation projects, which were established by Congress to provide power for pumping of water to supplement gravity-flow systems on reservation. These include Colorado River, Flathead, San Carlos, and the Wapato irrigation projects.

In addition to these formally designated projects, approximately half of the irrigated cropland in Indian Country is irrigated by tribal individuals or tribal government operators. Many of these systems are private ditch systems which retain the essential character and disposition of the original design. Because of the lack of formal funding for the operation, maintenance, and repair of these systems, some private systems are in disrepair. Nevertheless, several Tribal operations, such as Gila River, Navajo, Yankton, Winnebago, Standing Rock, and Lower Brule have fully operating and sophisticated irrigation systems.

During the course of development of irrigation in Indian Country, there has been considerable controversy over the construction, payment, and repayment of construction costs associated with Indian irrigation projects. The controversy has greatly affected the condition of Indian irrigation projects today. Beginning in 1914, 20 years after the Dawes Allotment Act,[7] irrigation construction costs were deemed reimbursable to the federal government either by the Indians or non-Indian successors in interest. In 1921, these debts became a lien on the property. Because of the inability of Indians or their non-Indian successors to repay the government, many irrigation systems fell into disrepair and lands fell out of Indian ownership. Acts of Congress in 1928, 1933, and 1936 either

TABLE 3.10 Partial list of Indian Irrigation Projects Authorized by Statue

	Year	Statute	State
Blackfeet Project	1907	34 Stat. 1035	Montana
Coachella Valley	1950	64 Stat. 470	California
Colorado River Reservation	1935	49 Stat. 240	Arizona/California
Crow Indian Irrigation	1891	26 Stat. 1040	Montana
Flathead Project	1904	33 Stat. 365	Montana
Fort Hall Project	1894	28 Stat. 305	Indiana
Fort Peck Project	1908	35 Stat. 558	Montana
Middle Rio Grande Pueblos	1928	45 Stat. 383	New Mexico
Navajo Project	1962	70 Stat. 96	New Mexico
San Carlos Project	1924	43 Stat. 457	Arizona
Soboba Project	1970	84 Stat. 1465	California
Uintah Project	1906	34 Stat. 375	Utah
Vaiva Vo Project	1965	79 Stat. 1071	Arizona
Wapato Project	1904	33 Stat. 595	Washington
Wind River	1905	33 Stat. 1016	Wyoming

deferred payment of debts or canceled inappropriate debts or liens against Indian and non-Indian property. The inconsistency in funding has contributed to the current deteriorated condition of many Indian irrigation projects.

The Bureau of Indian Affairs (BIA) currently has the primary management responsibility for Indian irrigation projects, although some tribes have contracted this authority from the BIA using provisions of the 1973 Indian Self Determination and Education Assistance Act. BIA management of irrigation projects has been severely constrained by institutional problems, lack of funding, and the interplay between land laws, repayment requirements, and land ownership patterns.

A 1975 report to the Senate Committee on Interior and Insular Affairs on the status of construction of Indian irrigation projects documented the need for more than $200 million just to complete and rehabilitate the 71 Congressionally authorized Indian irrigation projects currently administered by the BIA (Report to the U.S. Senate Committee on Interior and Insular Affairs on the Construction of Indian Irrigation Porjects, 1975). Estimates of the costs for OM&R on private systems are not readily available. OM&R needs on formal projects could represent a substantial liability to the United States as trustee for Indian Tribes. As a result, policy decisions related to Indian irrigation projects and water resources may significantly affect water resources available to irrigated agriculture.

NOTES

1. This section draws extensively on the following sources: U.S. Department of Commerce (1987), Bajwa et al. (1992), Solley, et al. (1993), Boggus et al. (1993), and Gollehon et al. (1994).

2. It should be noted that irrigation runoff is some cases is responsible for creating and maintaining wetland habitats, and curtailment of irrigation may on occasion actually harm or eliminate such wetlands.

3. Numerous studies by federal agencies document these impacts. See, for example, Bowman, David. 1994. Instream Flow Recommendations for Central Platte River, Nebraska. U.S. Fish and Wildlife Service, Denver, Colorado. May 23, 1994; U.S. Fish and Wildlife Service. Final Recovery Implementation Program for Endangered Fish Species in the Upper Colorado River Basin. Denver,

Colorado., September 29, 1987; Department of Interior, Bureau of Reclamation. 1993. Operation of Glen Canyon Dam, Draft Environmental Impact Study. Washington, D.C. May; Northwest Power Planning Council. 1994. Columbia River Fish and Wildlife Program. Portland, Oregon, December; National Marine Fisheries Service, 1995. Proposed Recovery Program for Snake River Salmon, Washington, D.C., March.

4. In 1994 the USDA expenditures on conservation and related programs affecting agriculture were estimated as follows: Conservation Reserve Program, $3.5 billion; wetlands programs, $56 million; water quality programs, $212 million; and other conservation, $1.5 billion (USDA, Agricultural Resources and Environmental Indicators, 1994).

5. The Central Valley Project Improvement Act also includes requirements that water districts and individuals who use federally supplied water assume responsibility for control and management of drainage discharges in order to comply with federal and state water quality standards (Section 3405(c)).

6. Unpublished BIA preliminary estimates for 1994.

7. 25 USC 348.

REFERENCES

Bajwa, R. S., W. M. Crosswhite, J. E. Hostetler, and O. W. Wright. 1992. Agricultural Irrigation and Water Use, ERS/USDA (Agriculture Information Bulletin No. 638).

Beard, J. B. 1982. Turf Management for Golf Courses. Minneapolis, Minn.: Burgess Publishing Co.

Bird, J. W. 1987. Transferability of Indian water rights. J. Wat. Res. Plann. Mgt. 113.

Boggess, W., R. D. Lacewell, and D. Zilberman. 1993. Economics of water use in agriculture. In Agricultural and Environmental Resource Economics. G. Carlson, J. Miranowsiki, and D. Zilberman, eds. New York: Oxford University Press.

Bryant, K., and R. D. Lacewell. 1988. Adoption of Sprinkler Irrigation on the Texas High Plains: 1958 to 1984. Texas Agricultural Experiment Station, Department of Agricultural Economics DIR 88-1, 5P-1. College Station, Tex.

Bureau of Indian Affairs. 1975. Report to the United States Senate Committee on Interior and Insular Affairs on the Status of Construction of Indian Irrigation Projects.

Bureau of Reclamation. 1987. Assessment '87: A New Direction for the Bureau of Reclamation. Washington, D.C.: Bureau of Reclamation.

Carlson, G., D. Zilberman, and J. A. Miranowski. 1993. Agricultural and Environmental Resource Economics. New York: Oxford University Press.

Casterline, G., A. Diner, and D. Zilberman. 1989. The adoptions of modern irrigation technologies in the United States. In Free Trade and Agricultural Diversification. A. Schmidt, ed. London: Westview Press.

Coastal Zone Act Reauthorization Amendments. 1990. P.L. 101-508, Section 6201 et seq., 16 U.S.C. Section 1455b et seq. (1993 Supp.).

Colby, B. G. 1994. The economics of Indian water conflicts: Competing property rights, shifting distributions of risk and the role of the market in policy implementation. In Water Quantity and Quality Disputes and their Resolution. A. Dinar and E. Loehman, eds. Praeger Publishers, Greenwood Publishing Inc.

Council for Agricultural Science and Technology (CAST). 1988. Effective Use of Water in Irrigated Agriculture. Report No. 113.

Deason, J. P. 1982. The Federal Role and the Objectives of Indian Water Resources Development. Paper presented at the 1982 ASCE Specialty Conference, Water Resources Planning and Management Division, Lincoln, Neb.

Folk-Williams, J. 1985. What Indian Water Means to the West. Western Network, Santa Fe, N. Mex.

Foxworthy, B. L., and D. W. Moody. 1986. Water-Availability Issues: National Perspective on Surface Water Resources. National Water Summary 1985—Hydrologic Events and Surface Water Resources. U.S. Geological Survey, U.S. Dept. of the Interior. Water Supply Paper 2300. Pp. 51-68.

Frederick, K. D. 1988a. The future of irrigated agriculture. Forum for Applied Research and Public Policy 3(2):80-89.

Frederick, K. D. 1988b. Irrigation Under Stress. Resources No. 91. Washington, D.C.: Resources for the Future.

Frederick, K. D., and J. C. Hanson. 1982. Water for Western Agriculture. Washington D.C.: Resources for the Future. Pp. 24-35.

Gibbons, D. C. 1986. The Economic Value of Water. Washington D.C.: Resources for the Future.

Gollehon, N., M. Aillery, and W. Quinby. 1994. Water Use and Pricing in Agriculture. In Agricultural Resources and Environmental Indicators. U.S. Department of Agriculture, Economic Research Service, Natural Resources and Environment Division. Agricultural Handbook No. 705.

High Country News. 1995. No More Ignoring the Obvious: Idaho Sucks Itself Dry. Vol. 27, No.3, February 20, 1995.

Intergovernmental Panel on Climate Change. 1996. Climate Change 1995: The IPCC Second Assessment Report, Volume 1. The Science of Climate Change, Summary for Policymakers. New York: Cambridge University (in press).

Jones, R. L., and G. L. Rando. 1974. Golf Course Development. The Urban Land Institute Technical Bulletin 70. Washington D.C.: The Urban Land Institute.

Lord, J. M., Jr., C. Burt, and G. Thompson, eds. 1981. Selection of irrigation method. In Irrigation Challenges of the 80's. St. Joseph, Mich.: American Society of Agricultural Engineers.

MacKichan, K. A. 1951. Estimated Water Use in the United States, 1950. U.S. Geological Survey Circular 115.

MacKichan, K. A. 1957. Estimated Water Use in the United States, 1955. U.S. Geological Survey Circular 398.

MacKichan, K. A., and J. C. Kammerer. 1961. Estimated Use of Water in the United States, 1960. U.S. Geological Survey Circular 456.

McNeely, J. G., and R. D. Lacewell. 1978. Water Resource Uses and Issues in Texas. Texas Agricultural Experiment Station.

Morgan, R. M. 1993. Water and the Land—A History of American Irrigation. The Irrigation Association, Fairfax, Virginia. 208 pp.

Murray, C. R. 1968. Estimated Use of Water in the United States, 1965. U.S. Geological Survey Circular 556.

Murray, C. R., and E. B. Reeves. 1972. Estimated Use of Water in the United States, 1970. U.S.Geological Survey Circular 676.

Murray, C. R., and E. B. Reeves. 1977. Estimated Use of Water in the United States, 1975. U.S. Geological Survey Circular 765.

Nash, L. L., and P. H. Gleick. 1993. The Colorado River Basin and Climatic Change: The Sensitivity of Streamflow and Water Supply to Variations in Temperature and Precipitation. EPA 230-R-93-009. Washington, D.C.: U.S. Environmental Protection Agency.

National Research Council. 1989. Irrigation-Induced Water Quality Problems. Washington, D.C.: National Academy Press.

National Research Council. 1992a. Sustaining Our Water Resources, Tenth Anniversary Symposium. Water Science and Technology Board. Washington, D.C.: National Academy Press.

National Research Council. 1992b. Water Transfers in the West: Efficiency, Equity, and the Environment. Washington, D.C.: National Academy Press.

Office of Technology Assessment. 1983. Water Related Technology for Sustainable Agriculture in the U.S. Arid/Semiarid Lands. U.S. Congress OTA-F212. Washington, D.C.: U.S. Government Printing Office.

Prusa, J., and J. Beditz, eds. 1985. Golf Course Maintenance Report. National Golf Foundation, North Palm Beach, FL, and Golf Course Superintendents Association of America, Lawrence, KS.

Rosenberg, N. J., B. A. Kimball, P. Martin, and C. F. Cooper. 1990. From Climate and CO_2 Enrichment to Evapotranspiration.

San Joaquin Valley Drainage Program. 1990. Fish and Wildlife Resources and Agricultural Drainage in the San Joaquin Valley, California.

Schneider, S. H., P. H. Gleick, and L. O. Mearns. 1990. Prospects for climate change. In Climate Change and U.S. Water Resources. P. E. Waggoner, ed. New York: John Wiley & Sons.

Schroeder, C. B., and H. B. Sprague. 1994. Turf Management Handbook. Danville, Il: Interstate Publishers, Inc.

Solley, W. B., E. B. Chase, and W. B. Mann, IV. 1983. Estimated Use of Water in the United States in 1980. U.S. Geological Survey Circular 1001.

Solley, W. B., C. F. Merk, and R. R. Pierce. 1988. Estimated Use of Water in the United States in 1985. U.S. Geological Survey Circular 1004.

Solley, W. B., R. R. Pierce, and H. A. Perlman. 1993. Estimated Use of Water in the United States in 1990. U.S. Geological Survey Circular 1081.

Stavins R. 1983. Trading Conservation Investments for Water. Environmental Defense Fund. Berkeley, Calif.

Tanji, K. K. 1990. Agricultural salinity problems. In Agricultural Salinity Assessment and Management. K. Tanji, ed. Water Quality Technical Committee of the Irrigation and Drainage Division of the American Society of Civil Engineers. New York: ASCE.

Thomas, A. T. 1994. Water rights: Legal aspects and legal liability. In Wastewater Reuse for Golf Course Irrigation. Chelsea, Mass.: Lewis Publishers. Pp. 94-95.

Turfgrass Times. 1965. Turfgrass as a $4 billion industry. 1965 (vol. 1).

Turner, J. H., and C. L. Anderson. 1980. Planning for an Irrigation System, 2nd ed. Athens, Ga.: American Association for Vocational Instructional Material.

U.S. Department of Agriculture. 1986. Agricultural Resources: Cropland, Water, and Conservation, Economic Research Service. Pp. 21-22.

U.S. Department of Agriculture. 1993. RTD Updates: Irrigated Land in Farms (No.2). Economic Research Service.

U.S. Department of Agriculture. 1994. Agricultural Resources and Environmental Indicators. Washington, D.C. Pp. 14.

U.S. Department of Commerce. 1989. 1987 Census of Agriculture. Various volumes, Pp. 21-22. 1989.

U.S. Department of Commerce. 1994. 1992 Census of Agriculture. Various volumes.

U.S. Environmental Protection Agency. 1992. Managing Nonpoint Source Pollution: Final Report to Congress on Section 319 of the Clean Water Act. Office of Water. Washington, D.C.

U.S. Environmental Protection Agency. 1994. National Water Quality Inventory: 1992 Report to Congress. EPA 841-R-94-001.

U.S. Fish and Wildlife Service. 1992. An Overview of Irrigation Drainwater Techniques, Impacts on Fish and Wildlife Resources and Management Options. Washington, D.C.

van Schilfgaarde, J. 1990. Irrigated Agriculture. In Agricultural Salinity Assessment and Management. K. Tanji, ed. Water Quality Technical Committee of the Irrigation and Drainage Division of the American Society of Civil Engineers. New York: ASCE.

Wahl, R. W. 1989. Markets for Federal Water: Subsidies, Property Rights, and the Bureau of Reclamation. Washington, D.C.: Resources for the Future.

Watson, J. R., ed. 1978. Proceedings of Waste Water Conference. Chicago, Illinois, November 978. Sponsored by the American Society of Golf Course Architects Foundation, U.S. Golf Association, National Golf Foundation, and the Golf Course Superintendents Association. Toro Company.

Watson, J. R. 1985. Water resources in the United States in Turfgrass Water Conservation. Cooperative Extension, University of California, Publication No. 21405, 1985. Oakland, California.

Webb, W. P. 1931. The Great Plains. New York: Grosett and Dunlap. Pp. 237-238.

Wilcove, D. S., and M. J. Bean, eds. 1994. The Big Kill: Declining Biodiversity in America's Lakes and Rivers. New York: Environmental Defense Fund.

4

Forces of Change and Responses

The appearances and methods of irrigated agriculture are as varied as the geography, the climate, and the cultural backgrounds of the people who practice it. But across the nation, fundamental and potentially far-reaching changes are challenging some of the basic premises supporting the use of irrigation, at least as traditionally practiced. This chapter explores these changes and their effects on the future of irrigation.

The extraordinary expansion of the use of irrigation in this century reflected, in part, its economic value—it was the primary tool used to make possible the settlement and growth of the American West. The importance of irrigation prompted a number of national and state policies to support the use of irrigation. One such policy, originating from decisions made at a number of points in time, was that federally supplied water for irrigation should be subsidized; that is, irrigators should have to bear only a portion of the full costs of their use of water (Wahl, 1995). Another policy was that a large portion of the available water supply would be committed to irrigation. This was not necessarily a conscious choice to favor irrigation, but it was the inevitable result of western water law, where those who were first to establish claims to use water had priority over any subsequent claimants (Bates et al., 1993). Under these prior appropriation principles, common throughout the western states, water uses are determined through the act of asserting physical control over the resource, and irrigators were often among the first to meet the criteria. A third policy was that irrigation should be free from at least some of the controls that might have been applied to reduce its adverse environmental effects; this was, indirectly, a subsidy that transferred the environmental costs associated with irrigation from the individual farmer to society at large.

But times and the nation's needs change, and these policies and the laws based on them are now being reevaluated and modified. The extent to which irrigation has been favored in relation to other values and interests is being reconsidered, and important changes are occurring. Society's desire for a more equitable distribution of the full range of costs and benefits is a key forcing function of change. The future of irrigation will depend not only on the extent and ultimate nature of the changes, but also on the manner in which adjustments and adaptations occur (Wescoat, 1987).

PROFITABILITY: A KEY INFLUENCE

At the present time, most irrigation-related decisions depend on farmers' and investors' expectations as to the profitability of the activity and the benefits and costs of irrigated relative to dryland farming. The principal determinants of the profitability of irrigated agriculture are the following:

- the overall state of the agricultural economy and markets, especially the benefits and costs of irrigated relative to dryland farming;
- the availability of water and its cost to the farmer and to society;
- available technology and management skills;
- the costs of other agricultural inputs such as labor, capital, and energy;
- environmental concerns and regulations; and
- institutions that influence how water might be used and the opportunity costs of using water for irrigation.

State of the Agricultural Economy

Investments in farming depend most importantly on the state of the agricultural economy in a region and, to a lesser degree, nationally. The price that farmers receive for their crops is a critical determinant of the profitability of farming. The profitability of irrigation is particularly sensitive to the level of crop prices because both crop yields and production costs are typically higher for irrigated than for dryland farming. In the past, federal farm income and price support programs have helped insulate farmers from some of the uncertainties of market prices generated by the forces of supply and demand. These programs provided an important stimulus to investments in irrigation.

Availability and Cost of Water

The timely availability of water for irrigation is critical for achieving good crop yields in many areas of the United States. Irrigation, by providing control over the timing and quantity of water available to plants, increases yields and reduces weather-related risks. In arid areas, irrigation is essential to commercial crop production; in semiarid areas, irrigation enables growers to achieve much higher and more reliable crop yields and expands the types of crops that can be grown successfully. Even in humid areas, irrigation produces higher and more stable yields than dryland agriculture and can be an important hedge against drought.

The willingness and ability of a farmer to irrigate depends in large part on the price and availability of water.[1] Access to inexpensive water was critical to the development of existing irrigated lands. The earliest irrigation involved diverting surface waters to riparian fields that could be irrigated with gravity flows. Costs rose as investments in reservoirs, pumps, and canals were required to increase assured supplies and to move water to more distant lands. Federal subsidies provided through the Bureau of Reclamation insulated some farmers from some of these cost increases. Where inexpensive or subsidized surface water was not available, cheap energy and technical breakthroughs such as turbine centrifugal pumps and improved high-speed engines reduced pumping costs and contributed to the widespread use of ground water for irrigation starting in the 1950s. However, the high financial and environmental costs of developing new water supplies and the growing competition for existing supplies are critical factors affecting the future of irrigation.

Available Technology and Management Skills

The ability of farmers to respond to changing water supply and economic conditions and their opportunities to do so depend in part on management skills and available technologies. High costs, including labor costs, and limited supplies of water are major factors underlying the ongoing shift from flood and furrow to sprinkler and microirrigation systems that require less water. The successful implementation of these water-conserving systems, however, depends on a higher level of management skills.

Costs of Other Agricultural Inputs

The costs of labor, capital, energy, and other agricultural inputs influence the profitability of farming in general; the relative benefits and costs of dryland versus irrigated farming; and the relative advantages of alternative irrigation systems. For instance, the profitability of sprinkler and microirrigation systems, which are capital-intensive but labor-saving, is sensitive to interest and wage rates. When water must be pumped from considerable depths and is applied under pressure, energy costs are an important factor in the profitability of irrigation. As energy costs rise, water-saving and energy-saving irrigation systems become more attractive.

Environmental Concerns and Regulations

Irrigation developed largely outside of the influence of modern environmental legislation and concerns. Irrigators claimed and diverted water from streams and aquifers and disposed of their return flows with little concern for the impacts on the quality of water bodies or on other water users. The future, however, is likely to be very different. Environmental concerns and economic realities have already brought the development of large new irrigation projects to a virtual halt. And in some areas, existing agricultural water uses are being challenged because of their impacts on water quality and fish and wildlife habitat.

Institutions

The future of irrigation also will depend on the institutions that influence the allocation of scarce water supplies among competing uses. Irrigators control many of the highest priority water rights in the West. In the past the demand to use the water for nonagricultural purposes has been relatively small. Institutional constraints on water transfers tended to keep already developed water in agricultural use. However, nonagricultural water demands are rising, and institutions for transferring water to other uses are developing. Consequently, irrigators are likely to have more and increasingly profitable opportunities to sell water for nonagricultural uses.

UNDERSTANDING THE RELATION BETWEEN FORCES OF CHANGE AND RESPONSES TO CHANGE

The forces at work to cause change and the responses to change are dynamic, interactive, and complex. To show that this is not a linear relationship and explore the nature of these processes, the committee developed a simple, illustrative matrix showing key forces of change and areas of response (Figure 4.1). Of course, a two-dimensional tool cannot adequately capture the complexity of the processes, but it can convey the basic principles at work. In Figure 4.1, major factors influencing irrigation are organized into three categories: changes related to the demands on and availability of water, economic changes, and changes resulting from concerns about environmental protection. In turn, responses to these changes are discussed under three headings: responses within the irrigation community; scientific and technological responses; and institutional responses. These "forces" and "responses" do not describe completely the current status and emerging trends in irrigation, but they do appear to be the most significant factors in evaluating change within irrigation.

Forces of Change	Response Areas		
	Irrigation Community	Science and Technology	Institutions
Related to Water			
Related to Economy			
Related to Environment			

FIGURE 4.1 Matrix of forces of change and responses.

FORCES OF CHANGE

The principal factors affecting the extent, nature, and profitability of irrigation are undergoing considerable change. These changes place pressure on irrigation to respond if it is to remain an important means by which agriculture and landscaping are to exist in many parts of the United States.

Changes Related to Water

Withdrawals and Consumption

Irrigation and livestock uses account for 82 percent of all consumption of water in the United States (Solley et al., 1993). Moreover, irrigation of lawns, parks, road landscaping, and golf courses accounts for much of the public municipal use in many areas of the country. In the western United States, withdrawals for agricultural use represent more like 80 percent of the total withdrawals and approximately 90 percent of total consumptive use. In short, irrigation is the dominant economic use of the nation's water supply.

That dominance is gradually eroding. In 1950, irrigation accounted for approximately half of all water withdrawals (Solley et al., 1988). By 1990, its share of total withdrawals declined to 40 percent. Although irrigation withdrawals during this period generally were increasing, other withdrawals such as for urban and industrial uses were increasing even more rapidly.

Historically, new demands have been met by developing additional water supplies through the construction of dams and interbasin conveyance facilities as well as ground water wells. Opportunities for such development increasingly are limited, primarily because financial and environmental consequences make the remaining potential sites less desirable. Reduction of ground water levels and aquifer storage in some areas limits additional development in these areas. Consequently, there is increased interest both in the reallocation of some of the developed water, particularly from agriculture, to new uses and also in the more efficient use of existing supplies.

Value and Cost

The use of large quantities of water for irrigation has been made possible, in part, by the low cost of that water. Consider that most irrigators pay less than one one-hundredth of a cent per gallon of water, some even much less than that.[2] The cost of water is a function of the cost of developing and making the water available. There is no charge for the use of the water itself. As mentioned, the costs of much surface water development—particularly for federally supplied water—have been substantially subsidized. Financing the rehabilitation, storage, diversion, and delivery systems at market rates would cause the cost of water to increase. Reduction or elimination of the federal subsidies for delivery of Reclamation project water would increase the cost of this source of supply as well. Otherwise, short of a governmentally imposed charge for the use of water itself or regulatory requirements imposing additional costs on the continued storage and use of water, there is little economic pressure on the cost of irrigation water from federal surface sources.

Ground water pumping, on the other hand, is greatly influenced by the energy costs associated with that pumping. Moreover, the greater the lift the more costly it is to pump the water. The influence of these two factors is demonstrated in Figure 4.2.[3] This example shows that as energy prices increase and the level of the aquifer declines, the costs of pumping ground water in the Ogallala aquifer are increasing. Similarly, the marginal value product of ground water in the Texas High Plains was estimated to be $5.98 per acre-foot in 1969 (Beattie et al., 1978) and by 1977, with the sharp increase in energy prices during this time, the marginal value product had increased to $19.67 per acre-foot in nominal dollars (Beattie, 1981). These increases inevitably affect consumption, although the degree is affected by a variety of variables.

In addition to the increasing cost of water, there is the considerable disparity between the economic value of using water in the irrigation of pasture land and some types of crops and its value in other uses. Young (1984), for example, estimated that, while the value of water for growing fruits and some specialty crops is much higher, 90 percent of the water used for irrigated agriculture has a value of $30 per acre-foot or less. Other studies of the value of irrigation show enormous variation, based on both the type of crop and the region in which the crop is produced. Potatoes, vegetables, and fruits produce the highest values—estimated typically at several hundred dollars an acre-foot or more (Gibbons, 1986). Pasture, sorghum, alfalfa, soybeans, corn, barley, and wheat tend to return the lowest values—from $3 to $30 per acre-foot (Gibbons, 1986). Such marked differences suggest that water staying in irrigation is likely to shift to crops producing higher economic returns. Moreover, with cities now purchasing rights

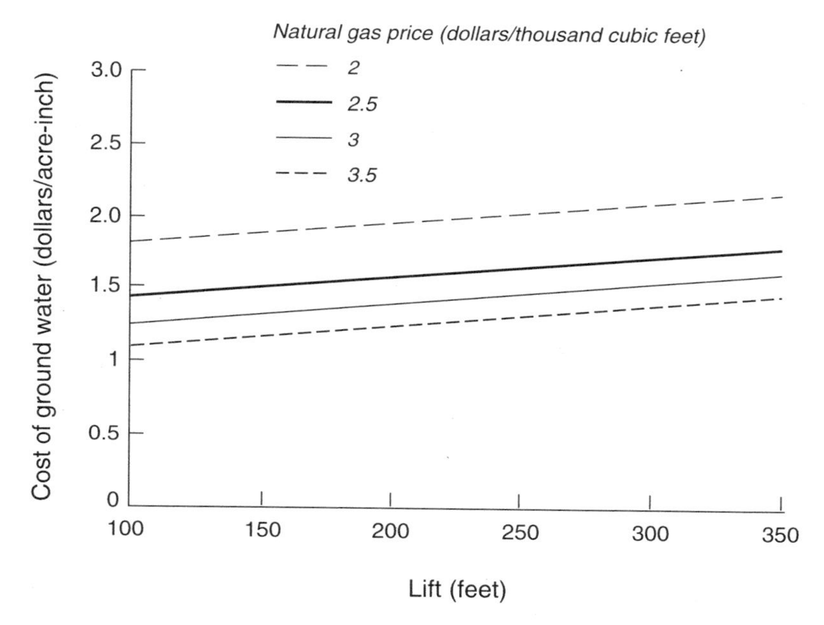

FIGURE 4.2 Ground water pumping is directly influenced by the price of energy and the distance the water must be lifted from beneath the soil surface. This example shows the relationship between lift and the cost of ground water for natural gas at various prices in the Ogallala aquifer; it assumes a sprinkler irrigation system operating at 45 PSI, pump and engine efficiency of 55 percent, and distribution efficiency of 75 percent. Source: Lee, 1987.

to the use of an acre-foot of water for $1,000 per year and more, implying an annual value of roughly at least $100 per acre-foot or more, it is reasonable to project that some irrigation water will shift to urban uses (some of which will be used for irrigation of urban landscaping) (National Research Council, 1992). In short, the changing nature of the values and uses of water are driving changes in the way water is used for irrigation purposes.

Indian Water Rights

One of the most significant potential forces of change is the settlement of American Indian water rights claims. Tribal water rights are rooted in the 1908 Supreme Court decision *Winters* v. *United States* (207 U.S.C. 564), which set out what has become known as the "Winters Doctrine." The Winters Doctrine provides that when the United States set aside land for a reservation, it implicitly

Box 4.1
Indian Water Rights

As competition for water has intensified in recent decades, the importance of settling uncertain Indian water rights claims has increased. In Arizona, for instance, 19 Indian reservations account for 20 million acres (28 percent) of the state's land base. Experts have estimated that the water entitlements of Arizona tribes, many of which remain to be quantified, may surpass the state's water supplies—most of which are already used by other parties. How did this come to be?

Under the doctrine of prior appropriation, the dominant mechanism by which most western states allocated surface waters, a water right is granted to those who first appropriate surface waters with the understanding that the right provides permanent access to the water as long as it is put to beneficial use on a regular basis. Later, more junior appropriators are only entitled to water not needed to satisfy senior rights.

In the 1800s, many Indian tribes entered into agreements with the U.S. government where they agreed to live on reservations. The principles of these reservations were outlined in treaties, which guaranteed that all tribal rights on a reservation were preserved except those expressly ceded away. Specifically, when these lands were set aside as reservations the natural resources were reserved for tribal people. In 1908, the U.S. Supreme Court confirmed this understanding in the historic "Winters decision." The court held that when the reservations were established, the United States implicitly reserved, along with the land, sufficient water to fulfill the purposes of the reservation. The court recognized these rights as having a priority date coinciding with the date that the reservation was established. And since reserved rights are not created by state law, they retain their validity and seniority regardless of whether tribes have put the water to beneficial use. Because Indian reservations generally were established prior to extensive western settlement, they hold extremely senior water priority dates.

For years, these senior rights had little practical value to tribes, and unexercised Winters rights posed little threat to other users. But in recent years tribes have begun to assert and develop their reserved rights, placing state water rights and Winters rights in competition. Settling these water rights claims is a tremendous challenge, and negotiations are underway in many areas, bringing many controversies.

SOURCE: Checchio and Colby, 1993.

reserved enough water to accomplish the purpose of the reservation, which is to provide a homeland for Indian people. The date of the water right is the date of the treaty between the tribe and the United States. In most instances, tribal water rights predate all other water users, and in the context of the prior appropriation doctrine are senior to all other users. The volume of water involved in settling Indian water rights claims will be important in shaping the future of the western United States, where secure access to water is the key to many economic activities.

The implementation of the tribal water rights, that is, the change from "paper water rights" to "wet water," has only recently begun in earnest. Tribal water use is now key to tribal economic development and is at the center of much of the current debate regarding the use, management, and development of water on major river systems such as the Colorado, Columbia, Snake, and Missouri.

In a 1963 case, *Arizona* v. *California*, the Supreme Court established the standard of measurement for an Indian water right as the amount of "practicably irrigable acreage" (PIA) on the reservation. As tribes enter the water rights adjudication process, calculations determining the quantity of water are based on a physical, economic, and technical evaluation of historic and proposed future irrigation projects for all of the reservation's PIA. Other water needs, including fisheries, wildlife, domestic, municipal, and industrial uses, usually add to the total tribal water claim.

The potential size of tribal water rights claims should not be underestimated. For example, water rights claims of the Missouri River basin tribes could total more than 19 million acre-feet, or approximately 40 percent of the average annual flow of the Missouri (Mai Sose, 1993). As of 1995, there are more than 60 cases in courts involving the resolution of Indian water rights claims. The total amount of water potentially involved in these claims ranges from 45 million to over 60 million acre-feet (Colby et al., 1992). As an alternative to litigation, the Department of Interior is actively engaged in 17 water rights settlement negotiations and is implementing another 13 settlements. Fewer than 10 of these efforts appear close to settlement. Table 4.1 presents settlements enacted in the last 10 years involving a total of 4.6 million acre feet of water (Colby et al., 1992).

Notwithstanding the PIA standard, recent national trends in the irrigation industry, the operations, maintenance, and replacement costs, and land tenure issues in Indian country continue to plague the use of Indian water for agriculture. Although tribes have expressed significant interest in water marketing, institutional barriers, state resistance, and the congressional authorization required for the interbasin and interstate marketing of Indian water remain as barriers to firmly identifying the amount of Indian water available for agricultural, instream, or other purposes.

In combination with tribal water rights, many tribes have treaty rights to instream flows for fishery resources, particularly in the Pacific Northwest. The quantification, exercise, and management of these rights may profoundly influence the future of irrigation by Indians and non-Indians alike.

A Changing Economy

The rapid and extensive development that occurred in the western part of the United States during the second half of the nineteenth century could not have happened without irrigation. Miners and the settlements that grew up in support of mining needed food. The only way crops can only be grown reliably in most

TABLE 4.1 Negotiated and Litigated Settlements of Tribal Water Rights

Settlement or Case/Date(s)	Indian Tribe	State	Quantity of Entitlement (Ac-Ft/Yr)	Comments
1. *Arizona* v. *California* Litigation/1963	Chemehuevi, Cocopah, Colorado, Fort Mohave, and Quechan (Ft. Yuma)	Arizona	905,496	• Upheld the Winters Doctrine • Established "practicable irrigable acreage" as a basis for quantifying Winters rights
2. Ak-chin Water Rights Settlement/1978, 1984, 1992	Ak-chin Indian Community	Arizona	85,000	• Original legislation modified due to impractical water supply plans • No local cost share; fully federally funded
3. Southern Arizona Water Rights Settlement (SAWRS)/1982, 1992	San Xavier and Schuk Toak Districts, Tohono O'Odham Nation	Arizona	66,000	• Allows limited off-reservation leasing • Provides federal projects water for tribe
4. Fort Peck–Montana Compact/1985	Assiniboine and Sioux Tribes	Montana	1,050,472	• Established permanent Compact Board to oversee implementation and to resolve disputes • Allows limited off-reservation leasing • Settlement never ratified by Congress
5. Salt River Pima–Maricopa Indian Community Water Rights Settlement/1988	Salt River Pima–Maricopa Indian Community	Arizona	122,400	• Complex multiparty water exchanges • Significant local cost sharing at insistence of federal government
6. Colorado Use Indian Water Rights Settlement/1988	Southern Ute and Ute Mountain Ute Tribes	Colorado	70,000	• Implementation delayed due to controversy over Animas–La Plata Project • Off-reservation water leasing provisions
7. San Luis Rey Indian Water Rights Settlement/1988	La Jolla, Rincon, San Pasquale, Pauma, and Pala Bands of Mission Indians	California	16,000	• No firm source of settlement water identified • Envisions conservation measures to provide water for entitlement
8. Big Horn Adjudication/1988, 1989, 1990, 1992	Wind River Shoshone and Arapahoe Tribes	Wyoming	500,717	• Expensive and contentious • Litigation ongoing • Only narrow issues addressed by courts; broad issues left unresolved
9. Truckee-Carson-Pyramid Lake Water Rights Settlement/1990	Pyramid Lake Paiute Tribes	Nevada	520,000	• Environmental concerns played major role in shaping settlement • Revised criteria for reservoir operation a key component

10. Fallon Paiute Shoshone Tribes Water Rights Settlement/1990	Fallon Paiute Shoshone tribes	Nevada	10,588	• Designed primarily to settle tribal claims against federal government for failure to construct a long-promised irrigation system • Not initially intended to settle reserved water rights claims
11. Fort Hall Indian Water Rights Settlement/1990	Shoshone and Bannock Tribes	Idaho	581,031	• Relies on unallocated federal water supplies • Authorizes establishment of a water bank • Established Intergovernmental Board to resolve settlement-related disputes
12. Fort McDowell Indian Community Water Rights Settlement/1990	Fort McDowell Indian Community	Arizona	36,350	• Considerable controversy over water supply • Secretary left to identify water source • Allows limited off-reservation leasing
13. Northern Cheyenne Indian Reserved Water Rights Settlement/1992	Northern Cheyenne Tribe	Montana	91,330	• Addresses issues of tribal/state jurisdiction and water rights administration • Most off-reservation leases subject to state law
14. San Carlos Apache Tribe Water Rights Settlement/1992	San Carlos Apache Tribe	Arizona	77,435	• Entitlement comprised primarily of Central Arizona Project water • Allows limited off-reservation leasing • Portion of water source strongly opposed by Arizona's non-Indian agricultural community
15. Utah Ute Indian Water Rights Settlement/1992	Northern Ute Tribe	Utah	481,000	• Designed primarily to resolve tribal claims against the federal government • No local cost share; fully federally funded
16. Jicarilla Apache Tribe Water Settlement/1992	Jicarilla Apache Tribe	New Mexico	40,000	• Strong opposition to out-of-state marketing provisions

NOTE: The figures listed in the quantity of entitlement column assume full implementation of all settlement provisions. For instance, the Pyramid Lake Paiute Tribe currently receives an average of 420,000 acre-feet/year and is not expected to receive its full entitlement of 520,000 acre-feet/year for another 10 to 20 years, when the fishery recovery plan established under the settlement is fully implemented. In addition, these figures represent *maximum* entitlements. For example, the Ak-Chin Indian Community is entitled to 85,000 acre-feet in wet years, but only 72,000 acre-feet during low flow conditions. These numbers generally represent allowed diversions, rather than overall depletions. For instance, the Northern Ute Tribe is entitled to divert 481,000 acre-ft/yr, but can only deplete overall streamflows by 248,000 acre-feet/year. Furthermore, these figures do not reflect additional tribal settlement benefits such as economic development funds or water delivery systems.

Source: Checchio and Colby, 1993.

parts of the West is with irrigation. Irrigated agriculture soon evolved beyond its role in support of mining and became a principal means by which the West itself could be settled. In this century, irrigated agriculture—at least in most locations—moved beyond its function as an agent for the settlement of the West to one of production farming. Today, irrigated agriculture is among the leading revenue-generating businesses in the western states. Moreover, agriculture in other parts of the United States is increasing its use of irrigation, including supplemental irrigation, and irrigation of lawns, parks, and other greenways in urban areas, particularly in more arid regions. By 2020 the world population will reach over 8 billion from the current 5.6 billion people. The increase in population must be matched with an increase in food and fiber production. The United States has the infrastructure to increase production and to market its increased production internationally. International marketing requires stable production which can be achieved by the expansion of irrigated agriculture, especially in the Southeast with high-value crops.

Other economic factors, however, run counter to this long-term expansion of irrigation. Agriculture itself is in the midst of some profound changes, with direct ramifications for the share that will remain irrigation-based. For instance, trading of agricultural products among nations in the global economy is an increasing practice. Although international trade will create potential new markets for products, it also will increase competition in domestic markets. Increased market competition puts pressure on farmers to grow relatively high-value crops as they look for ways to increase profits. But high-value crops often carry high risks as well and demand better management and business skills. Another apparent trend is a move toward vertical integration with a more corporate approach to production, processing, and marketing. In addition, the long-standing trend toward fewer and larger farms seems likely to continue as growers seek economies of scale.

The long-held views that agricultural production requires substantial governmental intervention and that the price of certain crops should be actively supported are increasingly being questioned. International trade agreements typically discourage governmental supports. More importantly, the public support for and financial capacity of the United States to continue farm support programs—at least at recent levels—are in doubt. Some crops grown with irrigation, such as corn, peanuts, cotton, and rice, are important beneficiaries of existing price supports. Moreover, lower-value, water-intensive crops, such as alfalfa, and pasture account for the lion's share of agricultural water consumption and receive subsidized water. These are policies that many find increasingly hard to justify. Zilberman (1994) has suggested that income and price support policies will matter less in the future, international markets will play a bigger role, and agricultural policies will be directed more to maintaining environmental quality. He also has suggested that many irrigators would prosper in a global setting because many of the higher-value crops that the United States is likely to have a relative advantage in producing are irrigated. Much irrigated agriculture will be in a position to compete effectively for limited water supplies and take advantage of technological advances. With conditions expected to be even more dynamic in the future, there will be opportunities for the innovative farmer, but also high economic risk.

Tribal water could be an important source of water for non-Indian irrigators. Where tribes have not yet put their water rights to use, this water often is used by non-Indian irrigators in the same basin. As tribal water rights are quantified through negotiation or litigation, tribes will retain considerable authority to determine the use and administration of tribal water on the reservation (Marx and Williams, 1995). While many tribes are not expected to develop new irrigated lands, the ability of tribes to market water could mean that non-Indian irrigators will have to compensate tribes for their continued use of tribal water. While tribal water marketing may not result in a reduction in water available for irrigation, it does enable the economic development of the tribal community through the generation of badly needed resources. American Indians and irrigators—historically not the best of friends—stand at the threshold of a new alliance that could be collaborative and mutually beneficial.

Changes Related to the Environment

As stated earlier, the principal environmental issues relevant to irrigation are those concerned with the protection and management of water supplies and water quality. In the last 25 years the public has become increasingly conscious of and concerned about environmental quality, endangered species, public health and safety, food safety, and the associated impacts of agricultural irrigation in these areas. Environmental issues related to water consumption and water quality in landscape irrigation are also beginning to receive greater attention. Policymakers and regulators at the federal and state levels have begun to respond to environmental concerns about irrigation, as evidenced by various efforts to control nonpoint sources of water pollution, water policies designed to protect instream flows, continuing restrictions on the types and application of agricultural pesticides, and other measures. Other environmental issues pertain to land use in agriculture generally—for instance, practices that diminish terrestrial habitats or otherwise impair habitat quality. Urban and suburban expansion into rural, agricultural regions has also given rise to conflicts over land use, waste disposal, recreational access, chemical use, and other issues. The relative significance of the numerous environmental issues faced by irrigators varies from region to region, but the nature of environmental issues confronting irrigation generally is the same coast to coast. The most serious problems exist in the West because of the large number of dams on originally free-flowing rivers and the amounts of water diverted.

The practice of irrigated agriculture has profoundly transformed the natural environment wherever it has occurred. Water development for irrigation has permanently altered aquatic ecosystems—rivers, lakes, streams, and wetlands—and the species that depend on them (Wilcox and Bean, 1994). Agricultural development accounted for 87 percent of all wetlands lost between 1950 and 1970 and 54 percent of those lost between the 1970s and mid-1980s (U.S. Department of Agriculture, 1994). Dam construction, water diversions, and ground water pumping for irrigation and other purposes have dewatered segments of some rivers, blocked the migration of anadromous fish, changed the natural hydrographs and temperatures of rivers, and damaged or destroyed riparian habitats. Runoff from agriculture including irrigation is now considered to be one of the largest sources of water pollution in rivers, lakes, and estuaries nationwide (U.S. Environmental Protection Agency, 1994).

At the same time, irrigation can provide important environmental benefits. With the construction of major dams, seasonal streamflows can be extended and water quality managed. Reservoirs that serve irrigation also provide flood control and recreational benefits, although there can be conflicts between reservoir operation for irrigation and for other uses. The growth of phreatophytes along ditches and the borders of irrigated fields may provide valuable habitat for wildlife as well as aesthetic benefits. Irrigated fields enhance ground water recharge and provide open space—true greenbelts during the summer growing season—often in contrast to the sparse growth on unwatered lands in many arid regions. The landscape industry provides aesthetic, recreational, and localized cooling benefits with golf courses, public landscaping, and private lawns.

In general, environmental policies and regulation affect irrigation by (1) restricting access to water resources, (2) restricting various activities that generate pollution or otherwise degrade environmental resource values, and (3) increasing the cost of doing business. Examples of relevant policy debates include continuing conflicts over implementation and reauthorization of the Clean Water Act and Endangered Species Act and debates over the need for direct regulation of agricultural nonpoint-source pollution (Young and Congdon, 1994). Still, irrigated agriculture has demonstrated a remarkable capacity to adapt to external stresses through changes in irrigation practices and technologies, scientific innovations, and economic diversity, and, to a limited extent, institutional change at the local and regional levels. The ultimate impact of environmental factors on irrigation will depend on how laws and policies are implemented and, more significantly, on how the agricultural community generally, and irrigators individually, respond to the environmental concerns of society at large.

Irrigators are likely to face more stringent requirements for protection of water quality in the future. The effect on irrigation of having to comply with water quality standards will be manifest as higher costs of water management and application (through the use of more efficient irrigation technologies and practices and improved irrigation scheduling). In some cases, farmers may choose to alter irrigation practices substantially or to retire lands that contribute significantly to pollution loads. The retirement of land as a pollution control option could be attractive where farmers are permitted by state law and/or federal law to market the unused portions of their water rights (Stavins and Willey, 1983). What these requirements will look like and how much flexibility irrigators will have in determining the appropriate pollution control options for their circumstances will be determined to some extent by the manner in which irrigated agriculture acts in shaping environmental policies and programs.

The landscape irrigation industry also contributes to water quality and quantity problems and may face increased regulations or other constraints as water pollution control measures are debated. Soil erosion and runoff during urban construction and the potential for leaching and runoff of nutrients and pesticides from established sites can lead to impacts on fish and wildlife habitats and aquatic systems generally through sediment, chemical, and thermal pollution of surface waters and pollution of ground water. These impacts are most likely to be addressed through the urban stormwater and combined sewer overflow provisions of the Clean Water Act.

The environmental issues confronting the irrigation industry have implications beyond questions of short-term economic return; they have to do with longer-term issues of sustainability. The ultimate effect of environmental regulation on irrigation will depend on the willingness of irrigators themselves to work with other interest groups and form new alliances in a changed political and economic context. Management options available to irrigators, such as changes in crops, investments in technologic improvements, water transfers, and conjunctive use, will be valuable for meeting environmental requirements and will enhance the sustainability of agriculture in the long term.

RESPONSES TO CHANGE

Forces of change are profoundly influencing irrigation. They are evoking responses that are shaping the future of irrigation in important ways. Many responses are positive—active steps taken to ameliorate problems and facilitate innovation. Some, however, are negative—resistance to change, whether shown by individuals, water agencies, or legislative bodies. As is to be expected, these forces sometimes conflict, with variable results. To gain perspective on the responses to change, the following sections explore the key response areas identified in the matrix provided in Figure 4.1.

The Irrigation Community

Irrigation in the United States has a rich history of developing internally initiated, innovative approaches to meet its needs. In the mid-to-late 1800s, for example, farmers and land developers in the western states organized themselves in a remarkable burst of creative, collective energy to construct water diversion and delivery systems, to create rules and procedures governing the use of water, and to manage the systems they had created. The irrigation systems and the mutual ditch companies and irrigation districts established at that time still exist, for the most part, and still provide irrigation water to millions of acres of cropland. The federal reclamation system, which arose in part because many private developments went bankrupt, became important in this century and made possible the expansion of irrigated agriculture throughout the West.

Maass and Anderson (1978) captured well the remarkable human ingenuity reflected in the early development of irrigated agriculture in the United States. There were obvious economies of scale to be gained by constructing a large central canal or ditch through which water initially would be diverted at an upstream point on a river and then contoured with the topography of the land to encompass as much irrigable land as possible. Water then could be delivered to these lands "under" the canal through branching ditches known as laterals. Some of the large canals were constructed by companies seeking to profit from the sale of lands made markedly more valuable because they could be irrigated. Most of the early irrigation systems in the western states, however, were constructed by entities created by collections of individual landowners intending to irrigate the land.

The challenges were many: inadequate financing, limited engineering capabilities, primitive earth-moving and other construction techniques, periodic floods that washed out diversion structures, highly variable flows of water that often were inadequate in the critical late summer months, relatively undeveloped legal rules governing rights to use water, and little or no enforcement of the rules that did exist.

Today the irrigation community faces challenges every bit as difficult and important as those encountered in the nineteenth century West. It is faced with a changing agricultural economy in which its economic position is less clear than at any time since perhaps the 1930s. It is faced with changing economies in some of the areas in which it has traditionally operated, changes that make irrigation a relatively less important part of the economic structure of those areas. At the same time, it is faced with almost certain reductions or even losses of some benefits it has enjoyed as a matter of public policy, such as subsidized crop prices and subsidized water.

There is little the irrigation sector can do by itself to influence the larger economic forces at work, but there is much it can do to effectively respond to those forces. As mentioned, the irrigated agricultural community is faced with much the same situation as is facing agriculture generally. Crop markets are becoming more competitive. Inevitably, such competition forces irrigation growers into a more businesslike approach to agriculture.

Much of irrigated agriculture already is operated in a highly efficient, businesslike manner, but economic pressures can be expected to accelerate this businesslike approach. Some growers will become processors and some will be major marketers. Crop selection will become even more price sensitive than in the past, and the pressure to keep costs down will continue to increase (Woolf et al., 1994).

It is tempting to predict that what might be called "lifestyle" irrigated agriculture—farming at a scale small enough to be operated largely or completely by family members, growing largely "staple" crops such as cotton, corn, alfalfa, or wheat, earning just enough to stay in farming—will not be able to survive in the changing economy. Almost certainly, some irrigators operating at the margin will not, and others will choose to leave irrigation for other reasons. There remains, however, a place in agriculture for this kind of farming. Despite the increasing technical and financial aspects of agribusiness, farming continues to be a way of life for many. It is one of the few means by which people can support themselves in a rural setting, especially in arid areas. As long as a living can be made, however modest, some will continue to pursue that option.

Competitive market pressures affecting traditional practices of irrigated agriculture also affect how irrigators look at their water supplies. Until recently at least, water has not been an especially costly input, and most irrigators have had little reason to think much about the economic advantages of using their water supplies differently—including conservation or leasing or selling the water to other users.

Particularly in places where the cost of water has increased measurably, irrigators are actively pursuing ways to use less water. In Texas, for example, the High Plains Underground Water Conservation District No. 1 instituted a series of programs beginning in the 1950s that have reduced the depletion of the Ogallala aquifer. Net depletion of ground-water-supplied irrigation of 5.5 million acres of land within a 15-county section of north Texas has decreased from an annual average of more than a million and a half acre-feet between 1965 and 1971 to an annual average net depletion of under 200,000 acre-feet between 1986 and 1991 (Wyatt, 1991). Initially, the district focused on persuading farmers to convert open ditches into pipes and to construct tailwater return systems. Beginning in 1978 the district initiated on-farm irrigation efficiency evaluations that included an analysis of well pumping efficiencies, an analysis of soil types and water holding characteristics, and an analysis of sprinkler system efficiencies. The process has been one of attempting to persuade irrigators to improve the efficiencies of their irrigation systems by demonstrating the economic benefits to be gained by doing so.

As the development of additional water supplies has become increasingly difficult, attention is turning to shifting some water from agricultural to urban, commercial, and industrial uses (National Research Council, 1992). The irrigated agricultural community generally has been uncomfortable with water marketing. There is a long tradition of regarding irrigation water as attached to the land—probably because agricultural use of the land otherwise would not be possible in many cases. This tradition is codified in many state water laws. Irrigators usually share water storage and delivery systems and have collective responsibility for the operation and maintenance of these systems. Trading or selling water within these systems has been commonplace. Taking water out of these systems for other uses in different locations threatens long-standing operational practices and raises questions about effects on the supply of those who remain within the system. The sale of irrigation water can bring the permanent loss of the associated agricultural activity, potentially affecting that part of the local economy dependent on agriculture.

New, more innovative approaches to water transfers are emerging that move water from agriculture to urban uses but with less harmful effects on a given agricultural economy (MacDonnell and Rice, 1994). For example, in 1989 the Metropolitan Water District of Southern California (MWD) paid for improvements to the water delivery system of the Imperial Irrigation District, in return for the use of the 100,000 acre-feet of water per year that those improvements are expected to save. In 1992, MWD entered into an arrangement with the Palo Verde Irrigation District for a land "fallowing" program that yielded 186,000 acre-feet of water in a 2-year period at a cost of $25 million (about $135 per acre-foot of water). Interested landowners within the district put together the package of lands that would participate in the fallowing program and, through the district board of directors, worked out satisfactory terms with MWD. In late 1994, agricultural and urban interests holding water delivery contracts from the California State Water Project entered into the "Monterey Agreement." Among other things, this agreement will open up the marketing of State Water Project water among the contractors.

Environmental concerns associated with water use also are affecting the irrigation community. For example, drainage from irrigation may contain contaminants carried from the soils such as selenium, as well as contaminants from pesticides and fertilizers (National Research Council, 1989). The Broadview Water District in the Central Valley of California developed an innovative program to reduce drainage water from its users and to better manage the drainage that is produced (Cone and Wichelns, 1993). District staff work directly with individual irrigators to encourage careful use of water and provide field-specific and crop-specific data describing water use. At the end of the irrigation season, the district brings farmers together to talk about successes and limitations. In 1989 the district instituted a tiered water pricing program under which the price of water increases as the quantity of water used increases. Water deliveries declined from an average of 2.88 acre-feet per acre in 1989 to 2.03 acre-feet per acre in 1992, and subsurface drain water declined from 4,626 acre-feet in 1986 to 854 acre-feet in 1992.

Creating incentives for environmental protection appears to be critical in enabling the irrigation community to respond effectively to new requirements. In the large region encompassing the Broadview Water District, district managers and irrigators in adjoining districts are responding to environmental concerns about drainage disposal by reviewing incentive-based options for reducing selenium and other contaminant loads to the San Joaquin River (Young and Congdon, 1994).

Scientific and Technological Responses

Technology has had a major role in the evolution of irrigation. Beginning in the nineteenth century, construction of diversions and canal distribution systems facilitated the growth of much of the area irrigated in the western United States. Irrigation in the 1800s was more art than science. Farmers learned by doing and shared what they learned with one another. But science and engineering soon came to play critical roles in irrigation. Perhaps the most significant technological innovations in this century were led by the Bureau of Reclamation, which designed and built water storage and delivery facilities throughout the West. At the time Hoover Dam was constructed in the 1930s, it represented a remarkable engineering achievement—one still greatly admired today.

After World War II, advances in technology came rapidly. The less expensive and readily available energy from the development of hydroelectric and natural gas supplies and the expansion of the electrical distribution network encouraged pumping. The advent of turbine pumps gave farmers access to ground water supplies and further increased the area irrigated. The technologies that spawned the additional irrigated areas will not, however, bring further increases. Ground water is being depleted, and only in limited areas is the development of wells bringing new land under irrigation.

Today, increased industrial and municipal water needs are being met in some instances with the storage and delivery systems that were first constructed for irrigation. The contemporary challenge, however, is not how to improve water storage and delivery; it is how to use water more efficiently. The technology of on-farm systems and improved management will come to the forefront in impacting the future of irrigation. Water must be used as effectively as possible to satisfy the increasing demands not only from industry and municipal users but also for the enhancement of fish and wildlife habitat. The need to reduce or eliminate water quality degradation requires new technology to improve or maintain the water quality in both surface and ground water.

Breeding has of course brought many advances, and there is hope that someday research will develop plants that use less water. Some successes have already been achieved with genotypes that mature early and avoid late-season drought, or develop deep root systems that gather large amounts of water. Proteins have been identified that may protect cells from death during severe dehydration, and the control of internal compounds for regulating stomata increasingly appears feasible. Some of the modern maize hybrids develop grain where others would fail during a drought, and wheat cultivars are available with increased drought tolerance compared to previous commercial types. Recent evidence shows that grain growth fails during a drought not because of a simple lack of the water necessary for the reproductive tissues but because the parent plant is unable to produce enough photosynthetic product to feed the developing grain. Thus, there is the possibility of altering the storage of photosynthetic products for use during dry spells.

Box 4.2
The Role of Molecular Biology in Breeding Crops for Improved Water Use Efficiency

Applied molecular biology is a nascent, but rapidly advancing discipline, which will in time effect the genetic improvement of crop plants. There are two principal approaches to using molecular biology to advance crop production, stability, and quality: (1) gene isolation and cloning and (2) genome mapping and marker-assisted selection.

The greatest advances in applied molecular biology have been achieved through targeting plant characteristics that are controlled by single genes and are easy to evaluate at the phenotypic level. Progress in gene isolation and cloning has been made toward improving disease, insect, and herbicide tolerance and resistance; enhancing biochemical composition (primary and secondary metabolites); and controlling breeding and phenology mechanisms (compatibility and ripening genes). However, although trait-based crop improvement for water-limited conditions may be analogous to breeding for disease and insect resistance, more complications arise because of the complexities of assessing crop characteristics for effect and the vagaries of plant-environment interactions.

An example of progress in breeding for improved water use efficiency has been accomplished through physiological-genetic studies of sorghum (*sorghum bicolor* (L.) Moench). Sorghum, a cereal grain, has evolved numerous mechanisms that have allowed it to become endemic and domesticated to semiarid, temperate, and tropical regions. Over the past 20 years through classical breeding, researchers have identified plant traits that when bred for can make sorghum survive in environments that experience intermittent or terminal water stress. These include (1) matching phenology to water supply; (2) osmotic adjustment of shoots and roots; (3) rooting depth and density; (4) early vigor; (5) increased leaf reflectance; (6) leaf

The development of tools of molecular genetics gives promise that understanding of plant-water relations will increase and be able to help minimize the use of water in crop production. The genetic control of deep rooting is being explored. These approaches may ultimately have application, but at the moment they are contributing mostly to our fundamental understanding of plant behavior. To hasten possible applications, specific target genes need to be identified that have a known function in preserving water while permitting plant growth. So far, the main barrier to progress has been the limited knowledge of which genes are important. Although promising, dramatic water savings from genetic engineering are not imminent. Therefore, for the moment, irrigation savings will need to be sought with existing crops that use water frugally and with improved efficiencies that decrease water delivery and application requirements.

area maintenance; (7) low lethal water status; (8) mobilization of pre-anthesis dry matter; and (9) transpiration efficiency.

Genome mapping and marker-assisted selection will be the approach of choice for molecular biologists working to support sorghum water use efficiency breeding efforts. More than likely, this strategy will be used for breeding for improved water use efficiency with other crop species as well. In complement with an aggressive mapping effort, reliable screening of traits (at the cellular, tissue, or whole-plant levels) is essential. High-resolution genome mapping can benefit improvement programs only if target characters are classified clearly and correctly.

Among the previously listed plant characters affecting water use efficiency in sorghum, mapping efforts have focused on osmotic adjustment because of potential impact and heritability of the trait. Osmotic adjustment is highly heritable and has been hypothesized to be controlled by two genes (one with recessive gene action on and one with additive). Based on initial linkage analysis, the gene exhibiting additive effects has been tagged by markers at 25 cM and 29 cM, respectively. However, to be useful for maker-assisted breeding, or, ultimately, gene isolation and cloning, much closer linkages must be found.

A complementary approach to current efforts that warrants consideration and may be a means to mobilize forces among all molecular biologists working in a plant family (e.g., grasses, legumes, crucifers) is to employ comparative mapping of traits. Based on evolutionary relationships, species in families frequently exhibit similarities among genomic maps and gene sequences. Therefore, insights, maps, and markers in one crop many benefit efforts in related species. For example, if a gene affecting water use efficiency is isolated in one crop, this target sequence may be of value to isolating and characterizing a similar gene in another crop. Comparative mapping has great potential for maximizing the positive impact of molecular biology across many crops.

Genetic Engineering

It is still probably too early to assess with accuracy either the potential or the limitations of genetic engineering for crop improvement (National Research Council, 1984). Gene transfer, for example, is unlikely to have a significant effect on agricultural production practices until the late 1990s. We are, however, on the brink of this time period but no significant breakthroughs on production with less water. The main benefit will be improved product quality and control of weeds and other pests, which will allow more economic return for the water used. Successes have been reported on engineering cotton plants that are tolerant of herbicides. When these new plants are cultivated, herbicides can be used to control weeds without damaging the crop. The Flavr Savr™ is a genetically engineered tomato that can be harvested ripe on the vine and brought to the market without softening. There are also biopesticides on the market that will control many pests without the use of chemicals and thus will improve the quality and production of food. Most of the anticipated advances are in the area of improving the characteristics of products for more economical uses and the control of pests and weeds.

Conserved Water

Although the actual volume of water that can be truly conserved and passed to other uses is, in general, small because conserved water is usually used elsewhere on the same farm, conservation is a widely accepted way to help meet increasing demands for water. However, the issues of cost and who should pay lend some uncertainty to how big an impact conservation can have for the future. Probably the most significant way to conserve water is by taking land out of production. Using improved application technologies (e.g., LEPA or drip systems) can also bring savings. In addition to freeing water for alternative uses, conservation can help limit the degradation of water quality. Increasing irrigation efficiency can reduce the amount of water diverted, but the return flows will be decreased and thus not available for downstream users. The removal of noneconomic vegetation can reduce water consumption, but the trade-off is loss of wildlife habitat and other environmental values.

Irrigation Systems

Improvements in surface irrigation fall into two general categories: improvements in the delivery of water to the farm and improvements in on-farm practices.

Storage and Delivery Systems The most significant changes in water delivery systems during this century are the incorporation of water-measuring devices such as metering turn-out gates and computerized flowmeters; the lining of porous earthen ditch systems with concrete and other impervious materials; the installation of "check" structures enabling better management of water in a canal or ditch; the use of reregulating ponds for the same purpose; the installation of debris collection systems; and the replacement of open ditches with pipes. All of these features tend to reduce the total quantity of water that must be diverted from a stream for delivery to farm headgates. Lining canals and installing pipelines can reduce the transmission losses. However, these "losses" act as a source of recharge for ground water that is used for irrigation elsewhere or that supports wetlands or other instream uses of water.

On-Farm Systems Surface irrigation systems such as flood and furrow systems are still the most widely used type of system. In areas with low-cost water, the typical surface irrigation system produces large quantities of runoff as the water flows across the field and infiltrates into the soil for use by the crop. Areas with more expensive water, such as the San Joaquin Valley, produce less runoff by using siphon tubes, which provide for a more uniform application into furrows of row crops. Land leveling, shorter furrow runs, and construction of borders and basins also provide more uniform irrigation. Gated pipe prevents losses in the distribution system and also provides for accurate control of the water to individual furrows.

The uniformity of infiltration into the soil is controlled by the time that water is on the surface and the characteristics of the soil. Because it takes time for irrigation water to move down the field, the upper part of the field generally receives more water than the lower end. To provide sufficient water to meet crop needs at the lower end of the field, more water than can be stored in the root zone often is applied to the upper parts of the field.

Sprinkler and microirrigation systems are designed to prevent surface runoff and apply the water uniformly to the entire field. Microirrigation applies water in a slow, precise manner. Water is delivered through a system of plastic tubes laid across or just under the surface and outfitted with special emitters designed to drip water into the soil at a rate close to the water consumption rate of the plant. Thus, rather than relying on maintaining soil moisture within the plant's root zone, drip systems seek to provide essentially a continuous supply of water (and other nutrients) directly to the plant. The range of on-farm efficiencies of alternative systems can be as great as 65 to 90 percent independent of the type of system (Keller and Bliesner, 1990). Closed, on-farm distribution systems are more readily adapted to automation, which is also beneficial considering the trend to larger farms and less-available labor.

A more uniform irrigation can facilitate more uniform crop growth and enhance crop production. According to the World Bank, under optimal management conditions, yield increasesof 20 percent or more have been reported per unit area utilizing drip irrigation, and of 40 percent or more per unit volume of water (Hillel, 1987). Microsprayer technology and low-head bubbler systems are more recent developments of microirrigation that offer advantages in some cases.

The major advantage of converting to sprinkler and microirrigation systems is the ability to more effectively achieve uniform water applications. When these systems are used, less water is needed at the farm gate or from ground water sources. In addition, farmers gain better control of inputs and water savings. Microirrigation systems can help reduce ground water withdrawals from overdrafted basins.

However, the increase in irrigation efficiency also reduces the amount of excess water that may have supported other uses. In the arid West, deep percolation is necessary to leach salts below the crop root zone. Without periodic leaching, the soil becomes saline and significantly reduces crop production. The reduction in runoff and deep percolation may affect downstream water users and irrigators who pump from the shallow aquifer. At the same time, the water not delivered will either stay in the stream or in the ground water reservoir. Each irrigation system is unique, and a detailed analysis must be made for each

Box 4.3
Irrigation Efficiency

There are many definitions of irrigation efficiency. Each definition is useful to describe the efficiency of different elements of the irrigation process. For example, efficiencies are defined for reservoirs, conveyance, basins, and on-farm application. But these varied definitions lead to some confusion. Many of the definitions would be better called performance parameters to reduce the false interpretation that an increase in the parameter makes more water available for other uses. An irrigation efficiency is often expressed as the ratio of the water used by the crop (in evapotranspiration, or ET) to that diverted to the field. The evapotranspiration for a crop is the sum of the transpiration by the crop and evaporation. To increase the efficiency, either the crop ET can be increased or the amount diverted decreased. Many factors influence crop ET:

Transpiration

Plants that undergo water stress will transpire less. For most crops a reduction in the transpiration causes a decrease in production. The yield reduction can be both in quantity and quality. Reducing the transpiration to save water is often not desirable since it can significantly reduce production. The better management option is to select different crops and plants which have a lower seasonal water requirement or to match periods of water availability. Transpiration is usually not affected by the irrigation method. With sprinkler irrigation, transpiration is decreased during the time that water is evaporating from the wet canopy surface.

Evaporation

The total ET will be increased with the addition of the free water evaporation. Evaporation from the soil surface and intercepted water on the leaves can be a significant part of the total ET. It is not possible to eliminate this use, but it can be minimized by not wetting the entire soil surface or the canopy. The frequency of

system as to how the hydrological cycle is affected with changes in diversion or pumping.

Irrigation Management

Careful management is critical to efficient water use. For instance, the scheduling of irrigations is a major factor in the amount of water actually supplied for crop production. If more water is applied than required, either evaporation, runoff, or drainage results. Poor scheduling can contribute to water degradation as excess water moves through the root zone, mobilizing salts and other constituents. Crops that are underirrigated will suffer water stress and reduced yields. The actual gains available through improved management will vary, of course, because the management skills of farmers vary and because some soils are particularly difficult to manage. In all cases, increased management brings higher costs, some real and some perceived. Thus irrigators must judge whether wetting contributes to the amount of evaporation. Sprinkler irrigation systems wet the entire soil surface and crop canopy that contributes to evaporation. Surface systems may or may not wet the entire soil surface contributing to soil evaporation. Microirrigation can leave part of the surface dry or can be buried and minimize the contribution to evaporation. However, there are many other considerations in selecting an irrigation system than just evaporation minimization, as illustrated by the questions below concerning water diverted.

Water Diverted

Diverted water is also an influence on irrigation efficiency. Water not used by the crop can follow various routes, depending on the specific site, including:

- Evaporates from reservoirs and canals and transpires from vegetation on the canal banks.
- Spills from the canal, where the timing of demands results in excess water in the canal.
- Is delivered to the farm headgate.
- Runs off from the field being irrigated and does not infiltrate into the root zone for use by the crop.
- In excess of the soil storage, percolates beneath the root zone and is not available to the crop.
- Is evaporated from the soil and crop canopy if wet during the irrigation process.

Water that evaporates is lost for further uses and will only return in the form of precipitation at an unknown location. Water that returns to the stream is available for another use. It moves to the ground water storage and can be pumped, or some may return to the surface streams. From a basin or regional perspective the water returning to the stream or ground water storage is available for other demands. In fact, many water rights and diversions are based on the return flow from water that was diverted for irrigation upstream.

the benefits exceed these costs. If the water is being reused, there may be no water savings from changes in the system and scheduling procedures. The economic benefits of upgrading and improving an existing irrigation system must be analyzed on a site-specific basis.

Current and Future Trends

A clear trend in irrigation today is the conversion of surface irrigation systems to more effective techniques such as sprinkler and microirrigation systems, especially where water costs or crop values are high. Center pivot, linear move systems, and other surface systems still irrigate approximately 55 percent of the nation's total irrigated area, but continued reduction in use is expected. Microirrigation now represents approximately 5 percent of the total irrigated area. Increases in the use of microirrigation systems are associated with high-value crops such as fruits and

Box 4.4
Opportunities to Improve Turfgrass Water Use Efficiency

Turfgrass is an integral part of the modern landscape—for aesthetic, recreational, and environmental reasons. As population has increased, combined with increases in leisure devoted to outside pursuits and in discretionary income, the turfgrass and landscape industry has become a major economic force—valued at $20 to $30 billion in 1992 (Watson et al., 1992). To remain healthy and vigorous throughout the growing season, turfgrass requires supplemental water—irrigation. Thus it has become a major water user and another competitor in the quest for limited water supplies.

As competition for water increases, it will be important for the turfgrass and landscape industry to find ways to reduce water use. Managers will need to find ways to conserve or recycle the water available. They will need to institute some fundamental changes in practice. Some opportunities to reduce water use in landscaping include the following:

- Acceptance of a lower level of quality for home lawns, recreational areas, some playgrounds, and general turfgrass areas maintained primarily for erosion control.
- Development and use of low-maintenance turfgrass, through either conventional breeding techniques or bioengineering, that require less fertilizer and less water to maintain acceptable quality and that are more stress tolerant.
- Expanded use of native grasses such as buffalograss (*Buchloe dactyloides*), blue gramma (*Bouteloua gracilis*), and curly mesquite (*Hilaria belingeri*), which are better adapted to adverse conditions.
- Technological advancements in the irrigation systems used to apply water to turfgrass, including the use of drip irrigation to supply trees, shrubs, and flowers in the landscape. Computerized controllers tied to on-site weather stations permit calculation of daily evapotranspiration (ET), and, at a predetermined time, the controller turns on the valve to replace that water lost to ET. Soil moisture sensors and rain shutoff switches also can conserve water.
- Improved public education on the importance of water conservation and how to irrigate most effectively.
- Increased use of nonpotable water—effluents and recycled water—for landscaping. Dual water systems for homes and commercial sites offer great potential to recycle water for use on landscaped areas.

vegetables. Limited water availability and high costs are the driving forces for these conversions. Lack of capital and an inability to pay for the investment with increased production in a short time period are still major constraints to conversions where water is plentiful and relatively inexpensive. In areas such as the high plains of Texas with limited ground water supplies, many center pivot and LEPA systems are being installed. California, Florida, and Hawaii have made significant conversions to microirrigation.

As we move toward the future, the competition for water and the need to increase efficiencies will continue to provide the driving force for conversion to sprinkler and microirrigation systems and the development of new technologies.

The 10 to 15 years required for widespread adoption of new technology must be recognized when developing policies for irrigation system improvements. Capital incentives can be provided to farmers to accelerate the conversion. Competing demands for water also will be met through reduction of irrigated area as urban growth and environmental concerns compete for a larger share of good-quality water. However, to the extent that farmers have capital to convert to more efficient irrigation systems, the total crop production can be expected to decline only slightly.

Researchers are investigating opportunities for plant adaptations that could reduce the plants' need for water. Improvements in water delivery and use technologies are already making major contributions to agriculture's ability to use less water to produce crops. Continued research is needed to develop strategies to prevent environmental degradation. At this point in time, the key limitations in improving agricultural water use efficiency are more economic and institutional than technical. The means for installing more efficient systems and the incentives to do so or to use more efficient practices simply have not always provided enough return to the grower to justify the expense of changing. The experience of this committee in talking to farmers reviewing the case examined in Chapter 4 indicates that the more successful farm operators are those who adopt new technology, strive for water efficiency, and manage capital-intensive operations.

Institutional Responses

There is a long history of highly developed institutions created to support the substantial infrastructure necessary for irrigation (Worster, 1985). Increasingly, institutions and their missions are adjusting to the forces of change already described. Table 4.2 provides a partial list of relevant institutions to illustrate the diversity of organizations acting at different levels. This section discusses some of the institutional responses at the federal, state, tribal, and local levels.

Federal Level

National policies affecting agriculture generally are in a period of transition. The conservation reserve program (CRP), for example, represents an important modification of the traditional farm-support programs because it supports farmers' incomes while reducing surplus production and promoting environmental values. The CRP pays farmers a yearly rental rate that averages about $50 per acre to remove land from production for 10 years (Faeth, 1995). The retired lands are supposed to be those that are highly erodible or that otherwise contribute to water quality problems. In 1993, more than 36 million acres of land were enrolled in this program. Nevertheless, critics note the inconsistency of farm programs that offer incentives to limit planted acreage and the production of specific crops, and reclamation policies that encourage irrigation and allow irrigators to grow these same crops with subsidized water (Moore and McGuckin, 1988).

Box 4.5
Irrigation Institutions Respond to Change:
Ground Water Management in Nebraska

The history of ground water management in Nebraska provides an example of how institutions can evolve to be more responsive to the changed availability of water. Prior to the early 1970s, neither ground water overdraft nor ground water quality was perceived as a significant problem, and Nebraska followed a laissez-faire management policy. State regulations were limited to well spacing requirements to prevent well interference and well registrations. A few single-purpose local districts existed to address specific problems such as runoff or drainage, but there were no comprehensive ground water management programs.

Rapid advances in irrigation development in the late 1960s, however, led to concerns about ground water quantity and to the recognition that small, single-purpose districts could not address the problem. In 1972, over 150 single-purpose resource districts were combined into 24 (23 as of 1992) comprehensive Natural Resource Districts (NRDs) covering the entire state. The NRDs are organized according to surface watersheds, governed by a locally elected board of directors, managed by a paid full-time employee, and funded by a property tax. They have broad responsibilities, including erosion and flood control, soil conservation, water supply, ground and surface water conservation, drainage, wildlife habitat management, recreation, and forest and range management. The NRDs have the authority to initiate ground water management controls to address both quantity and quality issues.

The establishment of the NRDs set the stage for other institutional adjustments—first the enactment of the Groundwater Management Act (GWMA) in 1975, followed by the Nebraska Groundwater Management and Protection Act (GMPA) in 1982. The GMPA authorized NRDs to establish ground water management areas and develop ground water management plans. Various amendments have strengthened the NRDs' authority to regulate ground water. The power of the approach is that it has given local authorities the capability to address ground water quality and quantity problems at the local level.

One example of an NRD responding effectively to a problem can be seen in the Upper Republican NRD, which when established in 1978 faced significant ground water depletion—levels had declined more than 25 feet in some areas, and continued declines were inevitable. When a water control area was first established in 1978, the area contained about 2,400 irrigation wells that were used to irrigate nearly 310,000 acres. Gradually, under the direction of the NRD, allocations were phased in, using incentives to encourage metering of well pumpage. Over time, they achieved a 34 percent reduction of pumpage per acre irrigated compared to the 1980 allocation. The regulations have affected irrigation and farming practices in the area; growth of irrigated area peaked in 1985 and has remained constant since. The rate of well installation has declined. Perhaps most importantly, although the water table has continued to decline, the rate of decline is much slower than it would have been without controls.

Another example can be seen in the Central Nebraska Natural Resource District. This NRD encompasses the heavily irrigated Central Platte Valley, where widespread irrigation began in the mid-1950s and developed rapidly in the 1970s. Over 80 percent of the cropland in the valley is irrigated, mostly via furrow irrigation but also increasingly with center pivot sprinklers. Irrigation wells are shallow and

Box 4.5 (continued)

high yielding. When nitrate levels were found to be two to three times the public health standard of 10 ppm, the NRD conducted research and determined that the problem was due primarily to excess application of nitrogen and irrigation water. In 1987 the district implemented a comprehensive ground water management plan that included provisions for reducing nitrate pollution. As the regulations have been phased in, environmental improvements have become evident. For example, the number of fields that were overfertilized by more than 60 pounds per acre has decreased from 26 to 14 percent, and residual soil nitrate has declined by 20 percent. Ground water quality in the most severe problem areas, in particular, has shown improvement. The historic trend of rising nitrate concentrations has stalled and begun to decline, in large part because of the NRD's efforts, and further improvements are expected as efforts to encourage expanded use of improved nitrogen and water management practices continue.

In the process of helping to develop the West's rivers and promoting irrigation, the U.S. Bureau of Reclamation (USBR) emerged as the supplier of one-fifth of all irrigation water in the United States, the country's sixth largest generator of electric power, and manager of 45 percent of the West's surface waters (Beard, 1994). But by the 1980s, the USBR was under stress. There had been no new authorizations for large federally financed irrigation projects since 1968, and the agency was being widely criticized for wasting federal funds, promoting inefficient water use, and damaging the environment. As a result, in 1987 the agency announced that it had accomplished its initial mission of helping to settle the West and that its mission was to change from one based on federally supported construction of irrigation projects to one based on resource management (Bureau of Reclamation, 1987).

It was another 5 years before there was much evidence that the agency did indeed view its mission as being broader than building dams and serving its traditional irrigation constituency. The 1992 Strategic Plan set forth new principles, goals, and strategies to guide the future Bureau of Reclamation. Its new mission is "To manage, develop, and protect water and related resources in an environmentally and economically sound manner in the interest of the American public" (Bureau of Reclamation, 1992). The agency's objectives now include providing a balanced approach to the stewardship of the West's scarce water and associated land and energy resources and diligently fulfilling its cost recovery responsibilities.

To fulfill these objectives, the *Blueprint for Reform: The Commissioner's Plan for Reinventing Reclamation* presented the following organizational principles for the USBR (Beard, 1993):

- Facilitate changes from current to new uses of water when such changes increase benefits to society and the environment;

TABLE 4.2 A Partial Listing of Government and Independent Agencies Concerned with Water Resources

Federal Level

Environmental Protection Agency
U.S. Army Corps of Engineers
Brueau of Reclamation
Bureau of Indian Affairs
Department of Agriculture (various divisions)
National Science Foundation
U.S. Geological Survey
U.S. Fish and Wildlife Service

Regional Level

Interstate Commerce Commission
Various river basin commissions
Great Lakes commissions
Various boundary water commissions
Resource conservation and development areas

State Level

Departments of Agriculture, Economic Development, Fish and Game, Public Safety, Health, Natural Resources, and Transportation
State engineers offices
Environmental quality boards
Pollution control agencies
Soil and water conservation boards
State planning boards and agencies
Water planning boards
Water resources boards

Tribal Level

Environmental quality boards
Natural resources commissions
Tribal water rights offices
Tribal government offices

Local Level

County agencies, boards, and committees
Municipal agencies
Township agencies
Drainage districts

- Emphasize the coordinated use and management of their existing facilities to improve the management of existing water and hydroelectric supplies;
- Encourage conservation and improvements in the efficiency of use of already developed water and hydroelectric supplies;
- Promote the sustainable use of the water and associated land resources in an environmentally sensitive manner;
- Facilitate integrated water resources management on a watershed basis;
- Conduct the agency in a fiscally responsible manner.

Box 4.6
Irrigation Institutions Respond to Change: Farmer-Interagency Collaboration in Central Arizona

In 1991, amid comments such as "why would I want to work with all the people who are making my life miserable?" and "I'm not being paid to give attention to [other agencies'] concerns," a coalition of federal, state, and county agencies, in cooperation with the Maricopa-Stanfield Irrigation and Drainage District (MSIDD) in central Arizona, initiated a demonstration project called the Management Improvement Program (MIP) to improve irrigated agriculture in the area. The MIP was initiated as a test to find ways to strengthen the local irrigated agricultural system through managerial and technological changes focused on improved management of natural resources and other inputs, leading to improved profitability, sustainability, and natural resources management. The program had three phases: in the diagnostic analysis phase, an interdisciplinary team of experts gathered data and reported nonjudgmentally on how irrigated agriculture was performing in the area; in the management planning phase, that understanding was shared with the community through structured activities involving the stakeholders; in the performance improvement phase, those plans were implemented and long-term mechanisms were established to sustain the effort after the formal end of the MIP.

When the test project ended 3 years later, evaluations found that the most immediate impacts were advancing a common understanding of the area's irrigated agriculture, identifying improvement opportunities, and improving communications and coordination among agencies and farmers. MIP efforts resulted in technology transfer and improved resource management. Further, the lessons are spreading as MIP participants, including both farmers and agency personnel, interact with farmers outside the original service area. The MIP was so successful that it left behind a local interagency Coordinating Group (CG), led by a farmer and with farmer members, to continue its work. Its mission is to provide areawide coordination of farmers and support services to serve the interests of agriculture while recognizing the importance of resource conservation. The CG sponsors town hall meetings to discuss important topics, a farmer-led program to exchange ideas on new farming practices, and other mechanisms to exchange information. The CG continues to serve as a way of providing focused, coordinated means to address the needs of the area's irrigated agriculture. To continue long-term, it will have to overcome challenges related to demands on members' time, personnel changes affecting the group and the agencies, identification and recruitment of leadership, and financial and other support needs.

It is too early to know just what the Bureau of Reclamation's restructuring and new mission will mean for irrigated agriculture. If the Central Valley Project Improvement Act (Title XXXIV of the Reclamation Projects Authorization and Adjustment Act of 1992) is taken as an example of future directions, it appears that reclamation policy is indeed changing from its tradition of promoting irrigation at the expense of the environment and the federal Treasury. California's Central Valley Project (CVP) is the largest federal water system in the country. Under the 1992 legislation, 800,000 acre-feet annually of CVP water that might otherwise have gone for agriculture is now dedicated for fish, wildlife, and habitat purposes. Furthermore, surcharges are being imposed on water users to finance environmentally related investments, marketing of federally supplied water is promoted, and tiered water pricing to encourage conservation is mandated. If this act provides a precedent for future legislation, many of the beneficiaries of past reclamation policies should expect to receive less federally supplied water in the future and pay more for what they do receive.

The focus of federal policies affecting water use has shifted sharply over the last 25 years or so toward greater protection of remaining streamflows and recovery of some of the environmental and recreational values that had been lost in the drive to provide homes, factories, and farms with water. This shift is evident in a number of legislative acts. The Wild and Scenic Rivers Act of 1968 precludes development activities that might significantly alter an area's natural amenities on thousands of miles of rivers and streams. The National Environmental Policy Act of 1970 requires all federal agencies to give full consideration to environmental effects in planning their programs. The Federal Water Pollution Control Act Amendments of 1972 (commonly known as the Clean Water Act), together with the Safe Drinking Water Act of 1974 and other legislation regulating the use and cleanup of toxic materials, have made water quality rather than water supply the driving force behind the nation's water-related investments. Requirements for protection of endangered species and their habitat under the Endangered Species Act of 1973 (ESA) are emerging as a major factor in some water management and investment decisions.

Agriculture has been a prime target in the debate over reauthorization of the Clean Water Act because changes in farming practices increasingly are viewed as critical to achieving further improvements in water quality. Past efforts to improve the quality of the nation's rivers, lakes, and estuaries have focused on controlling municipal and industrial point-source pollutants. These efforts are encountering high costs and diminishing returns in their ability to improve the quality of these water bodies to a fully usable condition. So far, agriculture has avoided the types of controls placed on the municipal and industrial point-source pollutants because the diffuse nature of most agricultural pollutants makes them difficult to control. Initially, the Environmental Protection Agency regarded discrete return flows from irrigated agriculture as point sources (Getches et al., 1991). Congress excluded agriculture from point-source regulation in 1977, and, since then, implementation of the Clean Water Act has not differentiated between dryland and irrigated agriculture. But this could change in areas where irrigation is a major contributor to water quality problems. Proposals for more deliberate regulation and enforcement of irrigation drainage, such as water quality standards in the San Joaquin Valley of California, where high selenium levels were deforming and killing migratory birds, provide a precedent for further regulation of irrigation return flows (Young and Congdon, 1994).

In 1937 the Soil Conservation Service (SCS) (now the Natural Resource Conservation Service) was created to assist farmers in preventing soil erosion.

The mission of the SCS was subsequently expanded to address soil erosion problems at a watershed level, as well as irrigation and municipal water storage. In the 1960s and 1970s, criticism of the SCS programs focused on impacts to fish and wildlife, loss of wetlands, and drainage problems, and its mission again was changed. The 1977 Soil and Water Resources Conservation Act required National Resource Inventories as the basis for SCS activities to reduce soil erosion, improve water management, reduce upstream flood damage, improve range condition, and improve water quality. This trend extended to the Food Security Act of 1985, which included a strong conservation title designed to protect wetlands. Today, the NRCS, along with other USDA agencies, is actively involved in providing financial, technical, and research services to farmers to conserve and protect highly erodible and environmentally sensitive lands and water quality. These changes are attributed to increasing competition over water resources, environmental concerns, and concerns for safe drinking water, recreation, and other public uses. The future mission of the NRCS is expected to be based on an ecosystem approach to resource planning to assist in meeting society's water needs and to protect, enhance, and restore natural resources (Carmack, 1994).

State, Tribal, and Local Levels

Individual states and tribes set the rules managing the water resources within their boundaries. Because of the importance of irrigation to the settlement of the western United States, state water law initially developed in ways that supported the needs of irrigators. The prior appropriation doctrine emphasizes the importance of seniority, a feature that is especially favorable to irrigators, who were generally the first to put large quantities of water to use in the West. It emphasizes physical control of water as a means of establishing a legal claim, another feature that favors

Box 4.7
Indian Water Marketing

Marketing of American Indian water rights by tribes is possibly one of the most important changes in water supply management and development. While many tribes have expressed interest in water marketing, the 10 tribes of the Colorado River basin, with a combined total of over 2 million acre-feet, are furthest along in discussions with affected states. The tribes contend that under *Sporhase* v. *Nebraska* (458 U.S. 941) tribal vested property rights to water can be marketed without regard to state and reservation boundaries. However, marketing of tribal water is generally subject to the approval of Congress. In the Colorado River basin, agreements with states could make marketing a reality. Marketing of Colorado River water would require an extraordinary commitment on the part of basin states and municipal water users, but would figure prominently in resolving the persistent water supply shortages. Other tribes with water resources that could potentially be marketed include all 28 Missouri River basin tribes, tribes in Washington state, and tribes in Oregon.

irrigation because water must be diverted out-of-stream to bring it to the fields for use. The seniority of tribal water rights now challenges the adequacy of most state water laws. In addition, the relationship between the tribes and the federal government usually requires federal government involvement.

At its base, water law is a system for allocating claims to use water; it is not designed to facilitate changes of those claims (MacDonnell, 1995). With the growing recognition of the need for reallocation of some existing water uses to new uses, states are moving to make their water laws and related review processes more able to accommodate voluntary transfers of water and water rights (MacDonnell et al., 1990). Similarly, the Bureau of Reclamation has made efforts to accommodate voluntary transfers of USBR-supplied water. The development of tribal water codes and management systems will add another dimension to the network of institutional structures related to water management.

One especially promising mechanism for facilitating both temporary and permanent water transfers is the water bank (MacDonnell, 1996). A water bank can be defined as "an institutionalized process specifically designed to facilitate the transfer of developed water to new uses." The potential effectiveness of water banks is illustrated by the successful use of this mechanism in California during the drought years of 1991, 1992, and 1994. A water bank can operate at a state, regional, or local level. It can be designed specifically to meet the needs of interested parties. One attraction for holders of water rights is the ability of a water bank to facilitate rentals and leases of water in addition to the more traditional approach involving the permanent sale of the water right. It offers the water right holder a choice about whether, in any given year, they would be better off renting or leasing water to another or using it themselves. It could provide water supply organizations such as irrigation districts and their water users a means of devising planned land fallowing schemes or other such approaches, similar to the arrangement involving the Palo Verde Irrigation District described above, and marketing the unused water without permanent reductions in its agricultural base or water rights holdings.

In addition to the matter of reallocation, western water law—with its emphasis on "use-it-or-lose-it"—remains in need of revision to provide incentives for more efficient water use (MacDonnell and Rice, 1994). Under the laws of most western states, irrigators installing more efficient irrigation systems lose the ability to legally use the portion of water that has been "saved." There is some logic for this: before the conservation strategy was applied, the "saved" water would have run off the land and subsequently been available to downstream users. But given the expense of installing more efficient technologies, one of the incentives that might encourage such action—namely, being able to make use of the additional water or to sell the water to another user—is lacking. The question is whether the saved water is the property of the one making the investment to use less water for a given purpose or whether it becomes the property of the remaining water rights holders within the water supply system.

The state of Washington has put in place an alternative approach, one in which government would pay for the improvements in return for legal control over the water no longer needed for irrigation. In 1994, Congress enacted a bill that could make funds available to help plan for and install more efficient irrigation systems in the Yakima Valley (MacDonnell et al., 1995). State and local funds also must be provided. Water saved through these improvements would then be managed by the State Department of Ecology.

A difficult problem is raised by the question of whether current state water law allows overuse of water. In theory, appropriative water rights are limited to the "beneficial" use of water. Thus, for example, Colorado defines beneficial use as "the use of that amount of water that is reasonable and appropriate under reasonably efficient practices to accomplish without waste the purpose for which the appropriation is lawfully made . . ." (Colorado Revised Statutes Section 37-92-103 (4).) In practice, the beneficial use standard has been very loosely applied (Shupe, 1982). It is instructive, for example, to compare the efficiency with which irrigators use water as a function of the seniority of their rights and the adequacy of their supply. Almost invariably, junior users are more efficient simply because they have to be.

The question of efficiency in water policy is a complicated one. Irrigation efficiency, for example, focuses on the amount of water used by crops for their evapotranspiration compared to the amount of water either diverted, delivered, or applied for this purpose (Keller and Keller, 1995). A modification of this traditional approach views efficiency as the relationship between the amount of water "reasonably and beneficially used" to the amount of water applied. The concept of "net" irrigation efficiency takes into account subsequent use of return flows, beneficial consumptive use, and nonbeneficial consumptive use. Still another concept proposed is the term "effective" irrigation efficiency—defined as the difference between "effective" inflow and "effective" outflow of water within a defined area. Like the net irrigation efficiency approach, this definition acknowledges return flows, but it also explicitly accounts for the need for some portion of the water supply to leach salts out of the root zone of crops.

None of these approaches to evaluating efficiency considers other related issues of the costs and benefits of the water uses that are being examined, nor do they permit consideration of the costs and benefits of making changes to increase the efficiency of use. Moreover, all of the approaches focus on irrigation use of water alone, without regard for other, nonirrigation uses nor overall watershed conditions. Thus, for example, even in the approaches that consider return flows there is no recognition that diverted water might have served valuable uses between the point of diversion and the point of return. Nor does it account for the possible adverse effects resulting from salts and other contaminants added to the water because of its diversion and use.

Whatever analytical approach is used, it is clear that there are few positive incentives for irrigators to make the investments necessary to reduce their water use. The laws of most states do not allow an irrigator to transfer "saved" water to another or to increase his or her consumptive use by, for example, using the saved water to irrigate additional lands (MacDonnell and Rice, 1994). Such restrictions are primarily intended to protect other downstream water users from possible adverse changes in their historical water supply. Oregon and Montana have addressed this concern by explicitly conditioning the ability to transfer saved water on the requirement that there be no injury to other water rights.

Given the absence of economic incentives, voluntary conservation efforts may not be sufficient. Consequently, some states and local water districts are turning to regulatory approaches to require more efficient water use. California has used its authority regarding "reasonable use" of water to require the Imperial Irrigation District to increase its water use efficiency (California State Water Resources Control Board, 1984). Oregon is proposing the institution of water conservation plans that would limit all uses to prescribed maximum amounts of water. Arizona is gradually reducing the allowable water duties for crops irrigated with ground water within described "active management areas" (MacDonnell and Rice, 1994). Tribal water rights settlements involving irrigation specify project water duties, efficiencies, and systems.

The critical decline in the level of the Ogallala aquifer in some areas has prompted regulatory responses at both the state and the local level. For example, well spacing requirements of some kind now exist for ground water development from the Ogallala aquifer in all of the states where it is found (Opie, 1993). Requirements for measuring withdrawals also now are common. In a few cases, users themselves have even imposed limits on the amounts of ground water that can be withdrawn beyond those provided in their original allocation (Kromm and White, 1992).

In addition to regulatory approaches, states and water districts are providing financial assistance as an incentive to implement water conservation practices. One form of such assistance is by providing low-interest loans to farmers to make soil or water conservation investments. The low-interest loans could be used to purchase distribution systems that are more technically efficient because of improved distribution efficiency or use of lower water pressure. Thus, theoretically, less water and much less energy could produce the same level of crop yield. However, with improved efficiency of water use and lower energy use, annual water use will not necessarily decrease because farms could use the conserved water to increase production on additional acreage. Irrigation water supply organizations have played a central role in the development of irrigation. As the needs shift from developing and delivering a water supply to solving a more complex set of problems, including pressure to ensure the continuing availability of water in a time of increasing competition and increasing concern about the environmental effects of irrigation, irrigation water supply organizations face different challenges. In many cases, these organizations are demonstrating real leadership in helping irrigation meet these challenges. In other cases, they seem caught in their more narrow traditional role and set to resist change rather than to facilitate it. They are key institutions with the potential to serve a critical function in ensuring that irrigation has a sustainable future. Their record to this point in serving this function, however, is mixed.

Ground water overdraft is one of many examples in which flawed institutions are delaying efforts to manage water resources effectively and to plan intelligently for the future. The mixture of water laws adopted by each state often depended on how arid the land was. Today, water law in the arid West protects senior users from supply interruptions and ensures that water entitlements will actually be employed, but efficiency is sacrificed. The prior appropriation doctrine and custom spell out an orderly way to allocate water resources, but they compromise the potential benefits of the resource through cumbersome treatment of water rights transfers.

Fortunately, state law also is changing to reflect increased interest in maintaining and protecting instream or in-place uses of water (MacDonnell and Rice, 1993). The long-held view that water should be managed almost exclusively for its out-of-stream uses, such as irrigation, is giving way to an increasingly widely held view that the ecological and recreational values of water are at least as important. To date, the changes in the laws of the western states regarding instream flows have had little direct effect on irrigation because the rights allocating water to irrigation use are very senior. Indirectly, however, attention to the in-place benefits of water highlights the massive manipulation of the rivers of the West that has occurred to facilitate irrigation. It raises questions, at a minimum, about whether there are ways in which existing irrigation needs can be met with less impairment of instream values.

Watershed-based approaches to water management are emerging in many areas, sometimes led by state, tribal, and federal agencies and sometimes driven by local interests (Natural Resources Law Center, 1995). Typically, these "watershed" initiatives are motivated by some overriding problem that is not being adequately addressed within the traditional legal and management structure. The watershed initiative institutes its own structure that includes the interests necessary to make desired change. Assuming agreement is reached on the nature of the change, the collective influence of the interests is used to produce the necessary institutional changes. Not uncommonly, traditional irrigation uses of water are a focus of these watershed efforts because these uses tend to dominate out-of-stream water use in many areas.

One well-known example of a watershed approach is Henry's Fork in Idaho, in which irrigation interests and others interested in improving and protecting streamflows in one of the premier trout fishing streams in the United States found sufficient common benefits to be able to work together with great success (Brown and Swenson, 1995). In the Yakima Basin of Washington, one of the leading irrigated agricultural areas in the country, agricultural interests spearheaded the formation of the Yakima River Watershed Council in 1994 (Farm Credit Services, 1995). This initiative was motivated by a recognition that the future viability of the agricultural economy in the basin depends on changes in the historical manner in which irrigation water is allocated and used (MacDonnell et al., 1995). Changes are needed both to enhance the treaty-based salmon and steelhead fisheries in the basin and to better meet existing and changing agricultural water uses. For the Columbia River system, the Columbia River Intertribal Fish Commission has proposed significant changes to the mix of irrigation, hydropower, and navigation operations with the goal of improving the condition and quantity of treaty-protected salmon stocks.

Irrigated agriculture has become increasingly separated from its urban-based neighbors. Watershed-based approaches to addressing changing water needs offer important opportunities for irrigation interests to connect in a more integrated way with emerging interests in other uses of water. They provide a vehicle for educating people about irrigation as well as for exploring ways in which agricultural needs for water can still be met while possibly providing benefits to other users. They provide a potentially important opportunity for irrigation water supply organizations to act positively in representing irrigation interests.

Such approaches are no panacea. They can be very time consuming, and their success often depends on intangible factors such as the personalities involved. They need to have the full commitment and participation of all key interests for their efforts to bear fruit. They often have funding and staffing problems, and they may be perceived as a threat by those representing traditional institutional interests. Nevertheless, watershed initiatives are taking hold in enough locations that they now represent a distinct and important approach to water management. They will be an important element in determining the future of irrigation.

CONCLUSION

In order to glimpse the future of irrigation in the face of competing demands for water, it is necessary first to identify and understand the forces of change affecting irrigators and how the farming community is responding. This chapter has addressed three forces of change—competition over water supplies, changing economic conditions, and environmental concerns—that appear to be the major determinants today in the practice of irrigation. Irrigators are responding to these factors in different ways and at different levels, ranging from the farm level, to the local or regional level (e.g., the irrigation district), to the level of state, federal, and tribal governments. Their responses are affected by developments in science and technology, adaptations within the agricultural community, and reforms in institutions and policies related to irrigation.

When these processes are examined at a national level, certain trends emerge and patterns repeat themselves, making it possible to glean a general understanding of the direction of change in irrigation, and possible irrigation "futures." The matrix presented in this chapter (Figure 4.1) provides a framework for examining and analyzing patterns of change and response, focusing on but a few of the myriad factors affecting irrigation. This tool is potentially useful for organizing a picture of a highly complex and dynamic activity.

At the same time, a limitation of the matrix is that it is static and therefore does not capture the dynamic nature of the pressure-response relationship. The seemingly independent factors that determine both the present and the future of irrigation are, in fact, interactive. Also, over time the adjustments irrigators make will have some feedback effect on the forces of change that caused them. For example, where environmental problems lead to adjustments by the agricultural community to mitigate them, that response may give rise to another pressure or factor for change. Finally, the matrix does not reflect the dimensions of time or spatial scale, which are key elements of sustainability. Some patterns of change and response may take place in a few years, whereas others last many decades. Similarly, these patterns are seen on individual farms, watersheds, or landscapes.

Although the forces of change and response described in this chapter are the most significant factors common to the future of irrigation nationwide, the matrix does not account for regional, cultural, and other differences within irrigation as a whole. Change occurs differently and to different degrees depending on the context in which it occurs. Responses are similarly site specific, varying according to the experience of and technology available to irrigators and the role and capacity of supporting institutions. Competition over developed surface water supplies occurs differently in California than in the Southeast and with different impacts (e.g., increased water prices, institutional changes, demands for new supplies). Whether irrigators respond by selling their rights, improving their irrigation efficiency, or turning to litigation depends on the context.

Chapter 5 presents a series of case studies to illustrate the major forces of change affecting irrigation in the agriculture and turfgrass sectors and how irrigators in different regions are responding. The case studies provide insights into many questions about the future of irrigation. For example, what are the issues and patterns of change common to irrigation throughout the country, and where do they vary among regions? What are the most significant forces shaping irrigation in a given region? How are irrigators responding? Are these responses of a short-term or long-term nature? Are they likely to significantly transform the industry, or are they merely an adjustment? Are some responses more "successful" than others? What are the most limiting factors for irrigation in the future? What opportunities for reform are suggested for the public and private sector institutions related to irrigation, and what should the role of these institutions be? Is the future implied by these changes and responses a sustainable one?

To date, agricultural irrigation has demonstrated a remarkable resilience and flexibility in response to changes in market conditions, pesticide and environmental regulations, conservation requirements, policy reforms, and even climate change. The net effect of current pressures on irrigation in the United States will depend in large part on how the industry responds and ultimately adapts to these changes.

NOTES

1. Price is the amount paid or charged for water in a transaction between two people and/or entities. Cost involves two components—all the financial outlays of individuals or entities necessary for water to be available (e.g., the costs of constructing and operating conveyance facilities) and other values foregone when the water is removed from its original use.

2. This figure assumes a cost of water of $30 per acre-foot or less, an amount that is on the high end of what most irrigators pay to use water. By comparison, the cost of urban water averages $1.66 per thousand gallons or 16/100 of a cent per gallon, which would include treatment and delivery system. (American Water Works Association, 1992, p. 79.)

3. Lee (1987) has calculated the cost of ground water in the Great Plains with the following equation: WC = 0.0014539*PNG*(Lift +(2.31*PSI)/(EFPMP*EFDS)), where WC = cost of pumping per acre-inch, Lift = feet from water table to surface, PSI = pressure requirement in pounds per square inch, PNG = price of natural gas in thousand cubic feet, EFPMP = pump engine efficiency, and EFDS = water distribution efficiency.

REFERENCES

Bates, S., D. Getches, L. MacDonnell, and C. Wilkinson. 1993. Searching Out the Headwaters: Change and Rediscovery in Western Water. Washington, D.C.: Island Press.

Beard, D. P. 1993. Blueprint for Reform: The Commissioner's Plan for Reinventing Reclamation. Bureau of Reclamation, Washington, D.C.

Beard, D. P. 1994. Remarks before the International Commission on Irrigation and Drainage, Varna, Bulgaria, May 18.

Beattie, B. R. 1981. Irrigated Agriculture and the Problems and Policy Alternatives. Western Journal of Agricultural Economics 7:289-299 (December).

Beattie, B. R., M. D. Frank, and R. D. Lacewell. 1978. The economic value of water in the western United States. In Proceedings of a Conference on Legal, Institutional, and Social Aspects of Irrigation and Drainage and Water Resource Planning and Management. New York: American Society of Civil Engineers.

Brown, J., and D. Swenson. 1995. The Henry's Fork: Finding mutual interest in the watershed. In Conference on Sustainable Use of the West's Water. Boulder, Colo.: Natural Resources Law Center.

Bureau of Reclamation. 1987. Assessment '87: A New Direction for the Bureau of Reclamation. Washington, D.C.: Bureau of Reclamation.

Bureau of Reclamation. 1992. Reclamation's Strategic Plan. Washington, D.C.: Bureau of Reclamation.

California State Water Resources Control Board. 1984. Misuse of Water by Imperial Irrigation District, Decision 1600. Sacramento, Calif.: California State Water Resources Control Board.

Carmack, W. J. 1994. Remarks for Workshop on the Future of Irrigation, Irvine, Calif., June 2.

Checchio, E., and B. G. Colby. 1993. Indian Water Rights: Negotiating the Future. Tucson, Ariz.: Water Resources Research Center.

Cone, D. G., and D. Wichelns. 1993. Responding to water quality problems through improved management of agricultural water. In Symposium on Water Organizations in a Changing West. Boulder, Colo.: Natural Resources Law Center.

Environmental Protection Agency. 1994. National Water Quality Inventory: 1992 Report to Congress. EPA 841-R-94-001. Washington, D.C.: Office of Wetlands, Oceans, and Watersheds.

Faeth, P. 1995. Growing Green: Enhancing the Economic and Environmental Performance of U.S. Agriculture. Washington D.C.: World Resources Institute.

Farm Credit Services. 1995. Yakima water users team up to resolve water issues, Yields Spokane, Wash. August.

Getches, D. H., L. MacDonnell, and T. Rice. 1991. Controlling Water Use: The Unfinished Business of Water Quality Protection. Boulder, Colo.: Natural Resources Law Center.

Gibbons, D. C. 1986. The Economic Value of Water. Washington, D.C.: Resources for the Future.

Hillel, D. 1987. The Efficient Use of Water in Irrigation. Technical Paper No. 64. New York: The World Bank.

Keller, J., and R. D. Bliesner. 1990. Sprinkle and Trickle Irrigation Design. New York: Van Nostrand Reinhold.

Keller, A. A., and J. Keller. 1995. Effective Efficiency: A Water Use Efficiency Concept for Allocating Freshwater Resources. Discussion Paper No. 22. Washington, D.C.: Center for Economic Policy Studies, Winrock International.

Kromm, D., and S. White. 1992. Ground Water Exploitation in the High Plains. Lawrence, Ks.: University Press of Kansas.

Lee, J. G. 1987. Risk implications of the transition to dryland agricultural production on the Texas High Plains. Ph.D. dissertation, Department of Agricultural Economics, Texas A & M University, College Station.

Maass, A., and R. Anderson. 1978. . . . And the Desert Shall Rejoice: Conflict, Growth, and Justice in Arid Environments. Cambridge, Mass.: MIT Press.

MacDonnell, L. 1995. Water banks: Untangling the gordian knot of western water. Rocky Mountain Mineral Law Institute 41:22.1-22.63.

MacDonnell, L. 1996. Managing Reclamation Facilities for Ecosystem Benefits. Boulder, Colo.: Natural Resources Law Center.

MacDonnell, L., and T. Rice, eds. 1993. Instream Flow Protection in the West, rev. ed. Boulder, Colo.: Natural Resources Law Center.

MacDonnell, L., and T. Rice. 1994. Moving agricultural water to cities: The search for smarter approaches. Hastings West-Northwest Journal 2:27-54.

MacDonnell, L., F. L. Brown, C. W. Howe, and T. A. Rice. 1990. The Water Transfer Process as a Management Option for Meeting Changing Water Demands. Boulder, Colo.: Natural Resources Law Center.

MacDonnell, L., R. Wahl, and B. Driver. 1991. Facilitating Voluntary Transfers of Bureau of Reclamation-Supplied Water. Boulder, Colo.: Natural Resources Law Center.

Marx, J., and S. M. Williams. 1995. Water Rights Administration on Indian Reservation. Proceedings, Albuquerque Conference, American Bar Association.

Moore, M. R., and C. A. McGuckin. 1988. Program crop production and federal irrigation water. In Agricultural Resources: Cropland, Water and Conservation Situation and Outlook. Report AR-12. Washington, D.C.: U.S. Department of Agriculture, Economic Research Service.

National Research Council. 1984. Genetic Engineering of Plants—Agricultural Research Opportunities and Policy Concerns. Board on Agriculture. Washington, D.C.: National Academy Press. Pp. 83.

National Research Council. 1989. Irrigation-Induced Water Quality Problems. Washington, D.C.: National Academy Press.

National Research Council. 1992. Water Transfers in the West, Efficiency, Equity, and the Environment. Washington, D.C.: National Academy Press.

Natural Resources Law Center. 1995. Watershed Sourcebook: Citizen-Initiated Solutions to Natural Resources Problems. Boulder, Colo.: Natural Resources Law Center.

Opie, J. 1993. Ogallala: Water for a Dry Land. Lincoln, Neb.: University of Nebraska Press.

Shupe, S. 1982. Waste in western water law: A blueprint for change. University of Oregon Law Review 61:483-510.

Solley, W. B., C. F. Merk, R. R. Pierce, and H. A. Perlamn. 1993. Estimated use of water in the United States in 1990. USGS Circular 1801.

Stavins, R., and Z. Willey. 1983. Trading Conservation Investments for Water: A Proposal for the Metropolitan Water District of Southern California to Obtain Additional Colorado River Water by Financing Water Conservation Investments for the Imperial Irrigation District. Berkeley, Calif.: Environmental Defense Fund.

U.S. Department of Agriculture. 1994. Agricultural Resources and Environmental Indicators. Economic Research Service, Natural Resources and Environmental Division, Agricultural Handbook 705. Washington, D.C.: U.S. Department of Agriculture.

Wahl, R. W. 1995. Natural Resources Subsidies. Washington, D.C.: Island Press.

Watson, J. R., H. E. Kaerwer, and D. P. Martin. 1992. The Turfgrass Industry. In Turfgrass. Waddington, Carrow, and Shearman, eds. Agronomy Monograph No. 32. Madison, Wis. American Society of Agronomy, Crop Science Socity of America, Soil Science Society of America.

Wescoat, J. L. 1987. The practical range of choice in water resources geography. Progress in Human Geography 11:41-59.

Wilcox, D. S., and M. J. Bean, eds. 1994. The Big Kill: Declining Biodiversity in America's Lakes and Rivers. New York: Environmental Defense Fund.

Woolf, S., B. Shepard, F. Peebles, C. Pintler, and J. Cofer. 1994. The on-farm perspective: Trends and challenges. Presntations at Workshop on the Future of Irrigation, National Research Council, Irvine, Calif., June 2-4.

Worster, D. 1985. Rivers of Empire: Water, Aridity and the Growth of the American West. New York: Pantheon.

Wyatt, A. W. 1991. Water management—southern High Plains of Texas. In Symposium on Innovation in Western Water Law and Management. Boulder, Colo.: Natural Resources Law Center.

Young, R. 1984. Local and regional economic impacts. In Water Scarcity: Impacts on Western Agriculture. Berkeley, Calif.: University of California Press.

Young, T. F., and C. H. Congdon. 1994. Plowing New Ground: Using Economic Incentives to Control Water Pollution from Agriculture. Oakland, Calif.: Environmental Defense Fund.

Zilberman, D. 1994. The effect of economics and agricultural policies on the future of irrigation. Presentation for the Workshop on the Future of Irrigation, National Research Council, Irvine, Calif., June 2-4.

5

The Irrigation Industry: Patterns of Change and Response

The productivity, profitability, and sustainability of irrigation in the United States are functions of numerous interdependent variables—physical, economic, political, environmental, and technological. These factors, taken alone and in combination, change over time and make the industry both diverse and dynamic. For this reason, it is impossible to depict a simple or homogeneous characterization of the irrigation industry in the United States.

Although it is possible to describe the nature of irrigation and the issues with which irrigators and the industry must contend in general terms, it is more difficult to generalize about the future of irrigation without looking at irrigation as practiced in different regions. Many of the key forces of change affecting irrigation vary in relative importance in different geographic regions. These factors also differ in relative importance between the agricultural and the turfgrass-landscape sectors of the industry. For example, while competition for water supplies and policies to protect environmental resources are issues affecting irrigation nationwide, the specifics of water supply problems and environmental restrictions are different in the Pacific Northwest than they are in the Texas High Plains. Policy reforms within the Bureau of Reclamation will have more significance for irrigators in the western states served by that institution than for irrigators in the southern and eastern United States. By the same token, the predominant environmental regulations affecting the turfgrass industry may not be of concern to agricultural irrigators. Within the irrigation industry, manufacturers of irrigation technologies do not face the same challenges and constraints as individuals who participate directly in irrigation activities.

Using a simple matrix, the preceding chapter described the relationships between forces of change and responses by the irrigation industry in the United States. This construct can be used to examine and analyze the experience of irrigators and supporting institutions and to formulate an overall picture of the industry, current trends, and the future of irrigation. This chapter presents four case studies to illustrate patterns of change and response as actually observed today. These case studies demonstrate how differences in conditions of water supply, concerns over environmental protection, and economic forces bring about varied responses. These trends can help identify the most significant pressures and provide insight into the magnitude and directions of change in the industry as a whole.

The case studies describe irrigation in four regions: the Great Plains, California, the Pacific Northwest, and Florida. The cases were chosen to illustrate a variety of irrigation patterns, processes, and problems. To aid in comparing these cases, it is useful to keep in mind several attributes that affect how irrigation is practiced in a given region. These are physical patterns, cultural patterns, functional relations, and jurisdictional relations.

- Physical Characteristics. The case study regions differ in terms of climate, hydrology, topography, and soils—factors that dictate certain irrigation practices, technology choices, public policy, and investments. For example, irrigation in semiarid regions, including much of California and the Pacific Northwest, depends on large-scale surface water delivery systems, most of which have been publicly financed and were built and operated by public agencies. Other regions, such as the Great Plains, are almost entirely dependent on privately developed ground water and have evolved pumping technologies and regional institutions to manage ground water. Humid conditions in Florida and the Southeast lead to different irrigation patterns.
- Cultural Characteristics. Cultural characteristics also differ significantly among regions and affect choices of irrigation technologies and practices, the structure and philosophy of local and regional irrigation institutions, and responses to environmental regulation and changing public policy. For example, American Indian irrigators operate in a markedly different cultural context than non-Indian irrigators, which is reflected in different philosophical, legal, and economic attributes. Individual tribes have strong spiritual values about water and land resources, values that influence their views about the political and economic value of those resources and how they are to be used. In addition, tribal resource management practices are oriented to long-term planning horizons (in contrast to the 50-year horizon commonly used by state and federal agencies). As sovereign nations, tribes have a fundamentally different relationship with federal and state agencies charged with management of water and other natural resources, and different policies and regulations pertaining to irrigation, reclamation, and crop production than non-Indian irrigation institutions organized under state laws.

Another example of cultural patterns can be seen in how different regions respond to technological and scientific innovations. For example, in California, the agricultural sector as a whole is characterized by a high average level of irrigation efficiency, but there are marked distinctions in irrigation efficiencies between farmers in different parts of the state. Different practices can be explained in part by physical and environmental parameters—the types of crops grown, soil characteristics, and climatic and hydrologic conditions. But some of the differences in irrigation efficiencies also are attributable to historical experience or family tradition and the irrigator's familiarity and comfort with new technologies.

Finally, cultural patterns also influence irrigators' perceptions of and responses to problems related to competition over water, environmental regulation, rising prices, and other factors. The types of conflicts that arise between irrigators and other interests, and how these conflicts are resolved, are uniquely a product of the cultural patterns that have developed over time.

- Functional Relations. Each irrigated area is defined by functional relations as well as physical and cultural characteristics. Although some irrigators grow crops for local and regional markets, others compete in global markets. Dairies tend to locate close to urban markets. The sites of processing plants influence crops grown in some regions. Many international markets are specialized (e.g., markets for mint from the Pacific Northwest), while other commodity markets are globally integrated (e.g., cotton and grains from the Great Plains). Some regions employ local and permanent labor, while others rely more on seasonal and immigrant workers. Crop subsidy programs target certain crops and will have a greater impact on growers in one region than another. All irrigated regions are interconnected by long distance financial markets and trade in irrigation equipment and supplies. These functional relations shape the economic geography of a region, just as climate and soils shape the physical geography.

- Jurisdictional Relations. All of the case studies depict relationships among political and administrative entities that define, to a greater or lesser extent, how irrigation develops; constraints on the availability of inputs; the context for solving environmental problems; and access to information, technical assistance, and technology. The California and Florida case studies, for instance, encompass multiple state agencies as well as overlapping jurisdictions of irrigation organizations and regional and local planning agencies. Additional jurisdictional levels are added in multistate cases such as the Pacific Northwest, where interstate, federal, and tribal responsibilities are considerable and policy goals are sometimes in conflict. The Great Plains case represents something of an exception to this rule because interstate water management policies, for surface and ground water, are relatively undeveloped. The Pacific Northwest and California cases involve, in different ways, international treaties, policies, and organizations. American Indian water rights, issues, and jurisdictional implications cut across regions, adding the dimensions of treaty rights and U.S. obligations.

The cases examined are complex. Each is a product of—and distinguished by—its physical, cultural, functional, and jurisdictional attributes. Each of the cases describes the character of irrigation in the region, the issues affecting irrigators, and how they are responding. In looking to these case studies for a picture of the future of irrigation, it is important to keep in mind that each case, while regionally or otherwise distinctive, is but a part of irrigation as a whole as practiced in the United States.

IRRIGATION IN THE GREAT PLAINS: TECHNOLOGICAL AND ECONOMIC CHANGES ASSOCIATED WITH DWINDLING GROUND WATER

The Great Plains marks the 100th meridian, the transition between the lush green of the East and the great desert of the West. Rainfall, which comes mostly in the summer, averages about 15 to 20 inches per year (Bittinger and Green, 1980). Precipitation varies greatly from year to year, and the area is classified as subhumid or semiarid. The climate, specifically the deficiency in rainfall, is the most significant characteristic in determining the Great Plain's environment and in making irrigation critical to the region.

Irrigation in the Great Plains depends almost entirely on the water in the Ogallala formation, a large aquifer system. In much of the Ogallala, the rate of withdrawal far exceeds recharge, which means that irrigators are in effect mining the ground water aquifer. Over time, ground water overdraft results in lower well yields, lower water tables, and increased pumping costs. Thus many irrigated areas of the Great Plains will face a transition as irrigation decreases and dryland production increases in its place. This prospect has serious implications for the primarily rural communities that depend on irrigated agriculture as their economic base and for the environment as land converts to dryland production and the threat of wind-driven dust increases.

The Great Plains region encompasses part or all of the states of Colorado, Kansas, Nebraska, New Mexico, Oklahoma, South Dakota, Texas, and Wyoming. Figure 5.1 shows the incidence of the Ogallala aquifer and its saturated thickness. Irrigation developed first in the southern region, and irrigated acres are now declining there. However, irrigated acreage is increasing in the northern part of the region.

Irrigation using ground water from the Ogallala developed after World War II as a result of the introduction of the centrifugal pump. The Ogallala covers 175,000 square miles (Zwingle, 1993). It sustains 20 percent of the irrigated acreage and provides 30 percent of all irrigation water pumped within the United States (Kromm and White, 1992b). The aquifer ranges in thickness from less than a foot to 1,300 feet, while averaging 200 feet (Zwingle, 1993).

The Ogallala contained an estimated 3 billion acre-feet of water before irrigation began. However, the Ogallala is a confined aquifer with an average

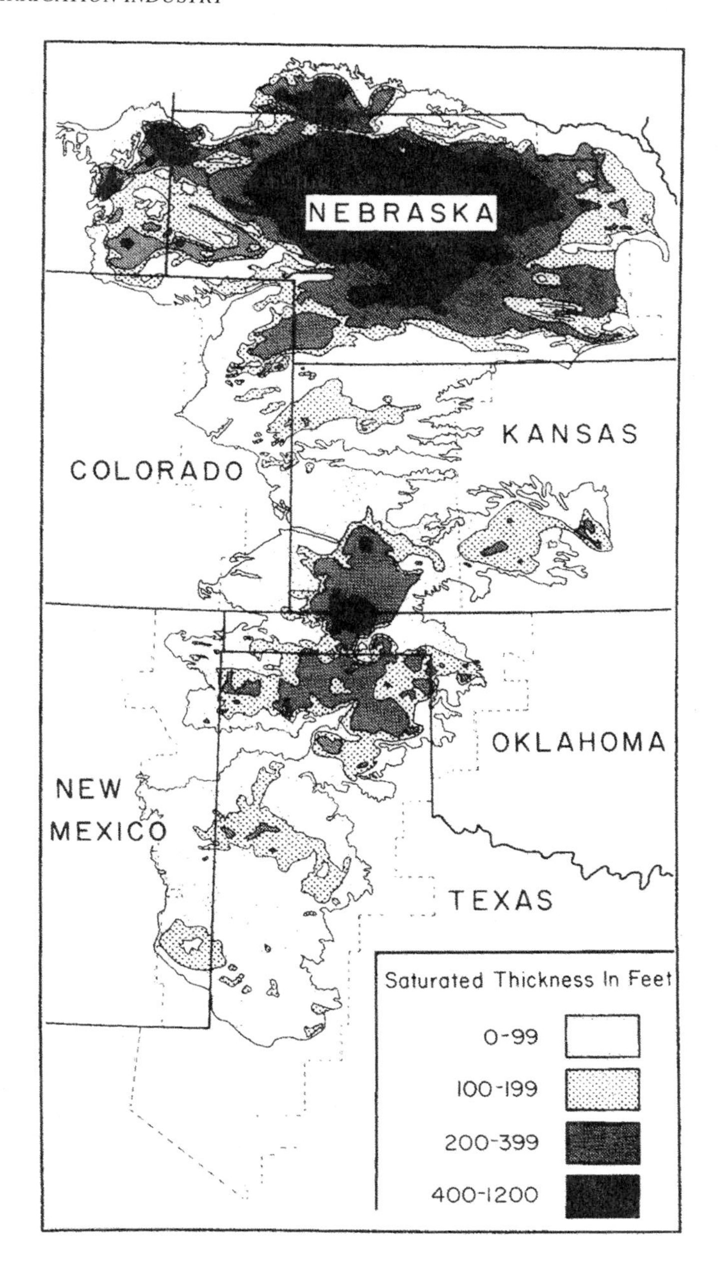

FIGURE 5.1 Saturated thickness of high plains aquifer, 1980. Source: Kromm and White, 1987.

recharge rate of about 0.5 inch per year; withdrawals, on the other hand, range from 1 to 5 feet per year. Even though there is a wide range in recharge rates, especially where there are sandhills, the Ogallala is being mined with withdrawals significantly exceeding recharge. Adjustments are already well underway to reduce water consumption. The critical issue affecting the future of irrigation in this region is the timing and types of adjustments that can be made and the effects these adjustments will have on agricultural crop production, total irrigated acreage, future rates of ground water withdrawal, and rural development.

Characteristics of Irrigation in the Great Plains

The major irrigated crops in the Great Plains are corn, wheat, grain, sorghum, soybeans, and cotton, with corn the dominant crop (Mapp, 1988). There are some high-value crops such as vegetables and sugar beets, but the acreage is very limited. Over 70 percent of the total value of crop production is from irrigated acreage (Beattie, 1981).

The extent of irrigated acreage in the different states of the Great Plains region is determined in large part by the incidence and characteristics of the Ogallala aquifer. Nebraska accounts for almost two-thirds (65 percent) of the annual pumping, with Texas using 12 percent, Kansas using 10 percent, Colorado using 4 percent, Oklahoma using 3.5 percent, and New Mexico, South Dakota, and Wyoming using less than 2 percent each. Over 87 percent of the aquifer is concentrated under Nebraska, Texas, and Kansas (Kromm and White, 1992b).

Irrigation across the Great Plains primarily relies on surface (flood) or sprinkler technology. Surface irrigation has moved from the open ditch and use of siphon tubes to closed delivery systems, use of shorter row lengths, and surge flow. Sprinkler systems include side roll, boom type, center pivot, traveling big gun, and linear move. In the last decade a large number of sprinkler systems replaced furrow systems, and LEPA (low-energy precision application) systems took the place of higher-pressure sprinkler systems (Bryant and Lacewell, 1988). Sprinkler-irrigated acres are increasing and by 1992 included 57 percent of all irrigated acres. Surface or flood irrigation was used on most of the remaining irrigated acres. Low-flow systems are insignificant in this region.

The pattern of irrigation development in the Great Plains region since 1959 includes some significant variations (See Table 5.1). The total number of irrigated acres increased to almost 13 million in 1978 but declined by about 20 percent in the following 9 years (Kromm and White, 1992a). Figure 5.2 shows total irrigated acreage across the Great Plains from 1959 to 1987. Most of the irrigated crops in the Great Plains are enrolled in the federal farm program. The total number of acres cultivated varies among the census years according to economic and weather factors.

The expansion in irrigated acreage is particularly significant in comparison to the change in nonirrigated acreage. Between 1959 and 1978 the average

TABLE 5.1 Total Irrigated Acres by State

State	1959		1969		1978		1987			
	Acres	Regional Percentage	Acres	Regional Percentage	Acres	Regional Percentage	Acres	Regional Percentage	% Change 1959-1987	% Change 1978-1987
Nebraska	1,937,036	28.1	2,620,382	28.5	5,046,815	39.1	4,967,607	47.8	+156.5	-1.6
Texas	3,921,189	56.9	4,379,471	47.6	4,496,514	34.8	2,616,446	25.2	33.3	-41.8
Kansas	548,642	8.0	1,195,548	13.0	1,956,087	15.1	1,607,301	15.5	+193.0	-17.8
Colorado	253,186	3.7	492,147	5.3	890,241	6.9	46,975	7.2	+195.0	-16.1
Oklahoma	53,342	0.8	259,647	2.8	264,155	2.0	246,367	2.4	+461.9	-6.7
New Mexico	226,435	3.3	253,456	2.8	12,923,331	2.1	209,728	2.0	-7.4	-22.2
Total	6,886,488		9,200,651				10,394,424		+50.9	-19.6

Note: Includes only acres included in the Ogllala aquifer, not total irrigated acres in a state.

Source: U.S. Census of Agriculture, 1959, 1969, 1978, and 1987 (Kromm and White, 1992a).

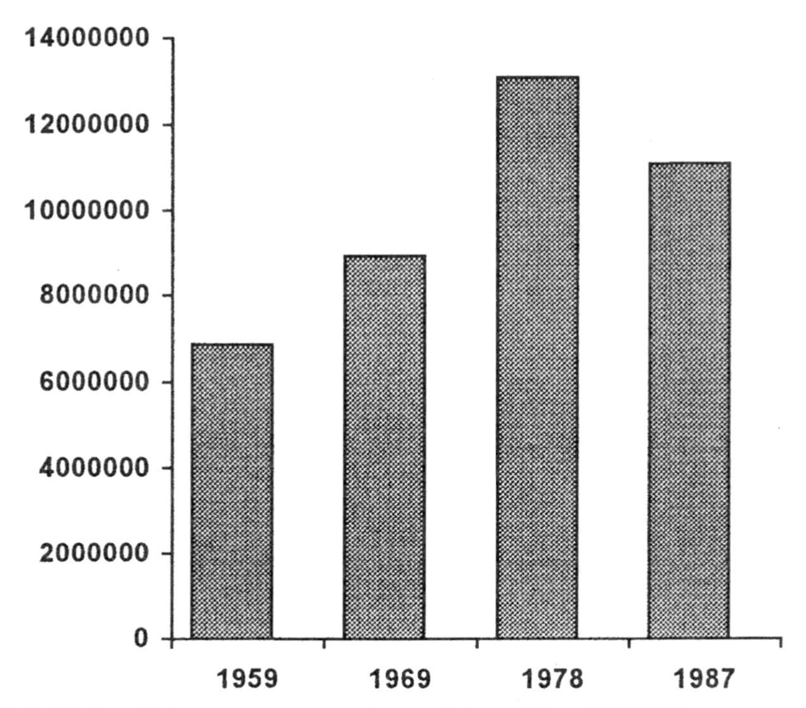

FIGURE 5.2 Total irrigated areas from the Ogallala aquifer, 1959-1987.
Source: U.S. Census of Agriculture (Kromm and White, 1992a, p. 24).

proportion of cropland under irrigation rose relative to nonirrigated acreage for the Ogallala in part of all six states. In Nebraska the proportion of irrigated acreage rose from about 28 percent in 1959 to almost 50 percent in 1987. From 1978 to 1987 the proportion of cultivated land irrigated in the Ogallala aquifer region declined in Texas and Kansas, increased in Nebraska and Colorado, and was about the same for Oklahoma and New Mexico (Kromm and White, 1992a).

The expansion in irrigated acres since the 1950s occurred with increased pumping of ground water. Ground water supplies will be the limiting factor in the development and distribution of irrigation for this region in the future. In 1978, some 12.9 million acres in the Great Plains region were irrigated with ground water. Projections for the year 2020 indicate that 5.4 million irrigated acres will revert to dryland farming or be abandoned (Banks et al., 1984). The areas where withdrawals can be expected to have the greatest impact by 2020 and beyond are New Mexico, Oklahoma, and Texas. These states account for over 3 million irrigated acres. Projections for Kansas and Texas show substantial decreases in irrigated acreage and corresponding increases in dryland acreage. Irrigated acreage in Colorado and New Mexico was projected to decrease without an accompanying increase in dryland acreage (Stewart and Harman, 1984). Nebraska is expected to continue to use 1.9 billion acre-feet of Ogallala water because of areas of substantial recharge and to be irrigating 11.9 million acres (Reisner, 1993). If these projections prove true, irrigated acreage in the Great Plains in 2020 and beyond will exceed current levels. However, the geographical distribution of irrigated lands will shift to northern states as southern areas adjust from full irrigation to supplemental irrigation to dryland production.

As ground water supplies continue to dwindle, particularly in the southern part of the Ogallala aquifer region, the transition to dryland will increase vulnerability to soil erosion from wind. The seriousness of wind erosion is shown by the 9 million acres enrolled in the Conservation Reserve Program (CRP) from the Great Plains. Erodible lands have been a priority since the 1930s dustbowl era, and under the CRP of the Department of Agriculture farmers receive payments to idle cropland and establish grass and other cover to reduce erosion. If the CRP is continued and gives priority on wind erosion control, it could be important in controlling wind erosion.

Forces of Change and Responses

In the Great Plains, as with the West generally, irrigation is most acutely affected by the rising cost of water. Agriculture, which accounts for about 88 percent of western water consumption, is not only the largest but also the marginal user of western water (Frederick and Hanson, 1982). Thus, as water supplies become more scarce, higher water costs threaten the continued expansion of irrigation as well as the continued production and profitability of current irrigators. In addition to ground water depletion and higher pumping costs, environmental concerns are putting more emphasis on water quality. These factors will play a significant role in determining the future of irrigation in the Great Plains, where some of the impacts and responses by farmers already are apparent.

Farmers over the Ogallala aquifer have been pumping water at a rate that exceeds recharge by severalfold (12 to 40 times more is pumped than is recharged). With recharge essentially negligible in most areas, continued mining of the aquifer will continue to reduce water availability, reduce well yields, and increase pumping lifts.

The impacts of increasing ground water depletion can be seen in the Texas High Plains, where annual pumping rates range from 5 to 8 million acre-feet, depending on prices and rainfall patterns (Lacewell and Lee, 1988). Continued pumping will result in a further decline of the water level in the Ogallala. A study done in the 1980s projected that the declining water table would support only about 55 percent of the 1980 irrigated acreage by the year 2000 and only 35 to 40 percent by 2030 (High Plains Associates, 1982). This same study for the six-state region forecasts that by 2020, water levels in the Ogallala will decline by 23 percent, with Texas having used two-thirds of its supply. At the same time, increasing lift and relatively expensive energy can be expected to maintain an upward pressure on the cost to pump. From the early 1970s to 1985, costs increased approximately 400 percent (Ellis et al., 1985).

Widespread water quality concerns have emerged with the development of irrigated agriculture in the Great Plains. A recent evaluation of the status of water quality and agriculture for the region (Lacewell et al., 1992) concluded that irrigated agriculture and confined livestock operations are the principal factors related to water quality problems across the Great Plains. Agricultural runoff is identified as the most extensive source of surface water quality degradation, accounting for about 60 to 80 percent of the water quality problems in the Great Plains. Soil erosion contributes to pollution through the combined effects of turbidity, siltation, and loading of nutrients adsorbed to the soil particles. Erosion in the Great Plains is dominated by wind action, which probably has a greater impact on soil fertility than off-site water quality.

A major source of ground water contamination is agricultural nutrients and pesticides. Ground water contamination by nutrients or pesticides has been documented in every state of the region except Wyoming, where contamination is suspected (U.S. Department of Agriculture, 1989). Of these contaminants, nitrogen fertilizers play a leading role because nitrates not used by plants are leached into the ground. One means for significantly reducing this pollution may be through the controlled application of water through fertilization and irrigation scheduling or "chemigation" (Kromm and White, 1992b).

Another nonpoint source of water contamination related to irrigation is runoff of pesticides and fertilizers into rivers, streams, and lakes. Across the Great Plains, farmers typically capture and concentrate runoff from irrigated fields in runoff pits, ponds, or playa lakes. Many farmers recirculate the water back through the irrigation system. Nevertheless, some runoff makes its way to other surface sources, and nutrients and some pesticides held in ponded water may lead to ground water contamination over the long term. A final cause of water impairment in the Great Plains is salinity. The relationship of salinity to other waste discharges is basically additive.

Current policies regarding agricultural nonpoint-source pollution encourage voluntary adoption of farming practices designed to protect surface water and ground water resources from agricultural chemicals and sediment. A major issue regarding policies directed to water quality in the Great Plains is the effectiveness of voluntary programs. Without significant improvements in water quality, there will be increasing pressure to adopt a regulatory approach to address agricultural nonpoint-pollution problems in the Great Plains and other irrigated regions of the United States (Lacewell et al., 1992).

The Ogallala experience shows that conventional farming with excessive water use cannot succeed over a long period of time and that adjustments toward more self-sufficient systems are needed. Some self-correcting mechanisms already exist that ensure that a given farming operation will require less water from the Ogallala each year. Because of higher pumping costs and lower well yields, farmers make adjustments in their farming organization, including the mix of inputs and equipment used. Farmers no longer feel that maximizing yield per acre is the most important goal; instead they have begun to concentrate on achieving an economically effective use of irrigation water. In the past decade, there have been adjustments in technology and agricultural practices, institutions, and rules and regulations. These adjustments have occurred at the farm level as well as at the regional level (Zwingle, 1993).

Conservation

Perhaps the most uncontroversial course of action for the Ogallala region is to conserve water primarily by increasing irrigation efficiency. As water costs rise, technologies and management practices that conserve both energy and water become more cost-effective and often essential to the continued profitability of irrigated farming.

In general, farmers in the Great Plains have a wide range of choices for responding to high energy and water costs before abandoning irrigation. These opportunities include improving pumping efficiency, installing tailwater reuse systems, reducing a sprinkler's operating pressure, institutions' irrigation scheduling, improving conveyance efficiency, monitoring soil moisture, shaping and leveling the land, irrigating alternate furrows, growing crops with lower water requirements or higher returns to water, and reducing the quantity of water delivered to a given crop. Other adjustments to increase irrigation efficiency include shortening row lengths for gravity-flow systems, converting to low-pressure sprinklers, and replacing worn sprinkler nozzles (Ellis et al., 1985).

Improved farming systems also contribute to Ogallala water conservation. Minimum tillage, rotating a row crop such as cotton or sorghum with wheat or other small grains, and careful use of herbicides for weed control to reduce the number of implement trips across the fields can cut costs and maximize the use of pumped and natural water. Another improved management practice is the limited irrigation-dryland system, in which the upper half of a field is fully irrigated, the next one-quarter is a tailwater runoff section using runoff from the fully irrigated section, and the lower quarter is managed as a dryland section solely dependent on rainfall. Throughout the Great Plains, this system offers a higher water use efficiency than full or conventional irrigation (Gilley and Fereres-Castiel, 1983).

Technologies for improving efficiency of water use in irrigation have made dramatic advances in recent years (Council for Agricultural Science and Technology, 1988). Improved management options for the effective use of irrigation water have become available through advances in irrigation equipment and have significant implications for the future of irrigation from the Ogallala. For example, advances in sprinkler systems include reducing the pressure to deliver water, including drop tubes to place the water in a furrow, and adding appliances to apply chemicals (multifunction systems).

However, there are some remaining disincentives to conservation. Legal institutions can discourage conservation in Kansas, where farmers who do not use their allotted water for 3 years lose the right to it (Zwingle, 1993). Without the appropriate incentives, technology also may encourage a greater use of limited ground water resources. As advanced techniques make better use of the water pumped, they lower its per unit cost and provide effectively more water during critical periods. Both effects encourage greater use of the limited supply.

Transition to Dryland Farming

The long-term result of aquifer mining, given the feasibility of other water supply options, will be a continuing shift to dryland farming. The transition to dryland farming will increase wind erosion (Lee, 1987). A transition from full irrigation will alter the structure of the agricultural industry. The returns to land (profit per acre) can be expected to decline which will result in falling land prices. Furthermore, those who can continue farming in the face of lower returns per acre will require larger farms, causing some displacement of current farmers. The Six-State High Plains Ogallala Aquifer Regional Resource Study concluded that, under conditions of crop prices and yield relationships of 1975 to 1980 and with currently projected rates of ground water depletion, a transition to dryland farming over the next 40 years would reduce gross farm income in the region by 25 to 50 percent (High Plains Associates, 1982).

The transition to dryland farming will have impacts for local and regional economies. The Great Plains has the largest concentration of farming-dependent counties in the nation. With lower levels of irrigation and dryland farming, there will be lower yields and less total crop output. This change will be reflected in a reduction in the demand for goods and services by production agriculture, which in turn will affect these local and regional economies. For the small communities, economic impacts may be particularly significant as decreased demands result in a higher per capita cost for services such as water supply, streets, hospitals, fire protection, and schools. Similarly, there will be decreased demands for supplies and other agriculture-related services including mechanics, input suppliers, fertilizer, seed, and fuel (Williford et al., 1976). The effects of a declining tax base already can be seen in small agricultural communities across the United States. The depletion of the Ogallala is likely to have serious socioeconomic implications for small towns in this region.

Irrigated acreage from the Ogallala is projected to decline in all states except Nebraska after 2020. The most serious decline in irrigated acres between now and 2020 has already begun and will continue in the southernmost part of the region (New Mexico and the southern Texas High Plains). A significant irrigated agriculture economy will remain across all of the Great Plains by 2020.

Adjustments in Technology

Future technological innovations will help farmers offset the impacts of higher water and energy prices. These include innovations in equipment, biotechnology, computer-based management systems, and other technologies.

A water-efficient irrigation system of the future may automatically schedule the quantity and timing of irrigation based on a computerized system that gathers and analyzes detailed information on soil water measured with a soil water sensor, climatic data for estimating evapotranspiration, and sensed crop response to the current environment. The irrigation system will also be used to apply fertilizers and pesticides. Global positioning systems (GPS) will be used in conjunction with sensors of water, salinity, crop fertility levels, and occurrence of weeds and pests to generate maps that will be processed with geographical information systems (GIS) to develop site-specific management recommendations. These site-specific management recommendations will be automatically transferred into computer instructions for controlling the irrigation systems for the timely application of water and chemicals.

The GPS will be used on harvesting equipment to provide yield variability maps, which can be used with other real-time information to determine the spatial variability and provide additional input to management information systems for subsequent crop management. Sprinkler and microirrigation systems that precisely control the application of water and chemicals will provide optimal production with minimal environmental hazards. Irrigation technology will truly become part of the information age.

Technological developments for increasing water supplies are far less promising. Experiments with artificial recharge to rebuild the aquifer have not resulted in any significant large-scale replenishment of the aquifer. The High Plains Underground Water Conservation District No. 1 in the southern High Plains of Texas has experimented with air injection, which sought to release water bound to sand particles in the desaturated zone. Although technically successful, the procedure is prohibitively expensive for the amount of water recovered (Opie, 1993). Cloud seeding is more scientific but equally unpromising in terms of cost (Opie, 1993). Even if a cost-effective weather modification technology existed, there would be major institutional obstacles to its adoption.

Regulations, Incentives, and Institutions

Voluntary conservation efforts may not be sufficient to protect the aquifer, and some regulations and economic incentives may be justified to manage the Ogallala water more efficiently. Various regulations have been introduced in the Ogallala region with the creation of local water districts. The Texas High Plains Underground Water Conservation District No. 1, the Oklahoma Water Resources Board, and the Southwest Kansas Groundwater Management District No. 3 were established with a common mission—although not always well served—to play an active role in managing and regulating the regional allocation of ground water. Management practices and regulations include issuing drilling permits, controlling well spacing, developing workable recharge, installing water meters, and preventing water waste. Today there are limitations in essentially all the states, which range from pumping limits to well spacing regulation. Parts of Kansas, Texas, and Nebraska now put various limits on the spacing between new wells (Zwingle, 1993). Nebraska is requiring meters on wells. Some towns in Kansas and Nebraska have made arrangements whereby the town uses fresh water and delivers wastewater to agriculture for use in irrigation.

State and local governments can make low-interest loans to farmers for investments in soil or water conservation. Texas has implemented a state-supported program to provide low-interest loans to farmers to purchase "water conserving irrigation equipment" or make other water-conserving investments (Lacewell and Segarra, 1993). The low-interest loans can be used to purchase distribution systems that improve distribution efficiency or use lower water pressure. However, with improved efficiency of water use and lower energy use, there is the potential that farmers will expand production to new acreage, thus eliminating any net benefit in water conservation. Research results do not support the contention that state-supported low interest loans for farmers to purchase more efficient irrigation equipment will necessarily extend the life of the Ogallala (Lacewell et al., 1985).

Numerous institutions are constraining opportunities to manage Great Plains water resources effectively and to plan intelligently for the future. The states of the Ogallala region (except Texas) rely exclusively on the prior appropriation doctrine of water law, which protects senior users from supply interruptions and ensures that water entitlements will be used but which does sacrifice efficiency. In particular, the prior appropriation doctrine and water use customs inhibit the potential for more efficient allocation of water through water marketing and water rights transfers. Texas follows the English doctrine that the owner of the surface owns the water beneath and may sell or lease their water. Furthermore, there are no constraints beyond ground water districts' rules to the drilling of new wells. This means senior water users are vulnerable to junior water users.

The government farm program has traditionally been an important component of a farmer's decision making. Changes in the 1995 or later Farm Bills could clearly have a large effect on irrigation decisions as well as cropping decisions. The trends in the globalization of world markets and other factors may bring unexpected changes or gradual adjustments in the Ogallala region.

Conclusion

Because of its vast Ogallala aquifer, the Great Plains has a distinct advantage over water-short parts of the country. Agriculture is by far the highest consumer, using 90 percent of Ogallala water, but there are relatively few metropolitan cities to compete for the water. Lubbock and Amarillo, Texas; Guymon, Oklahoma; and Liberal and Garden City, Kansas, are unlikely to become new Denvers, Phoenixes, or Tucsons. However, the continued heavy consumption of the non-renewable Ogallala water, no matter how judiciously regulated by state agencies, will eventually lead to shortages.

Most projections about the future of irrigation in the Great Plains, however, seem overly conservative. They assume that inputs such as irrigation water and fertilizer, as well as irrigation efficiencies, will remain fixed. They underestimate the ability of producers to adapt through changes in inputs and improvements in irrigation equipment, management practices, and the like.

In actuality, as water costs rise, farmers and supporting institutions are responding with more efficient irrigation and farming systems, alternative crop varieties requiring less water, improved knowledge of the relation between plant growth and water stress, and changes in laws and policies. Continuing changes in crop prices and technology (both equipment and management) also tend to offset or mediate effects of water shortages and rising costs. Most importantly, a smooth transition from water development to water conservation and reallocation will require fundamental changes in the long-held attitudes toward water use.

IRRIGATED AGRICULTURE IN CALIFORNIA: UNCERTAINTY AND CONFLICT IN THE FACE OF CHANGING DEMANDS

Irrigated agriculture in California is extensive and far from homogeneous, with distinctly different climatological, soil, and market opportunities among regions. These distinctions are reflected in the differences between growing regions and the diversity of commodities produced in the state. There are approximately 9 million acres of irrigated land in California. The state is a leading producer of 58 commodities, including fruits, nuts, and year-round vegetables. In 1992, farm receipts were valued at $18 billion; including multipliers, irrigated agriculture contributed $70 billion or approximately 10 percent to the state's economy (California Department of Water Resources, 1994).

Agriculture provides 365,000 farm employment jobs. There are roughly five jobs in agricultural input, processing, distribution, and so on, for every on-farm job. Overall, agricultural employment accounts for approximately 10 percent of the total jobs in the state (California Department of Water Resources, 1994; Rosenberg et al., 1993). California's farm receipts represent 10 percent of the total U.S. farm production. Until the early 1980s the amount of irrigated cropland in California was expanding. However, it appears that a number of factors will tend to reduce that acreage in the near future.

California is an arid state, and there would be no significant farming in the state without irrigation. Precipitation varies dramatically from north to south.

Seventy percent of the water supply exits in the northern one-third of the state, while 70 percent of the demand for water occurs in the southern portion of the state, south of the Sacramento–San Joaquin delta. Major water development projects were built beginning in the 1920s and 1930s and were spurred on during those years by extended droughts. Project construction continued through the 1950s and 1960s on the federal Central Valley Project and the State Water Project. By 1960 the total irrigated acreage in the state had increased from approximately 1 million acres in 1890 to 6.5 million acres.

In 1950, California officially declared there were ample water supplies, including its share to the Colorado River, to serve all future agricultural and urban uses. In time, this prediction proved overly optimistic. Proposals for the further expansion of water projects drew strong opposition beginning in about 1970 from environmentalists concerned about declining fish populations and degraded aquatic habitats. Construction of major features of the Central Valley Project and the State Water Project—Auburn Dam and the Peripheral Canal—has stopped. Major north coast rivers, including the Eel, Trinity, and Klamath, which constitute 40 percent of the state's water supply, were placed under both state and federal Wild and Scenic Rivers Acts and, thus, are protected from development. By the 1990s the situation in California had changed, with the state recognizing that the total of all future demands for water, including urban, agricultural, and environmental, will be greater than the developed supply and that it is unlikely that any new water supply projects will be constructed. This has created a new era of competition for water, which will tend to make water less available to agriculture than it has been in the past.

Environmental concerns over the preservation of fish and wildlife species and water quality have played a major role in transforming public water policy and management of water supply projects in California. Today, the key environmental issue is providing adequate flows for water quality and to protect fish and wildlife, including several endangered species, in the San Francisco Bay/Sacramento–San Joaquin delta estuary. Although environmental concerns are an important factor in determining the availability of water for irrigation in California, they are not the only factor affecting the future of irrigation. In the 1980s, rising costs of farm production inputs, including water and energy costs, soaring land prices, and high interest costs to finance both land purchases and annual production costs, slowed the rate of agricultural growth. Ground water overdraft is a persistent, and in some cases severe, problem facing irrigation, as are drainage and salinization problems.

California agriculture serves its state market, the national market, and the international market. It is not expected that California's position in these markets will change significantly in the future. However, following a decade of droughts, environmental confrontations, natural disasters, and economic downturns, it is clear that California needs to clarify its water management policies to resolve conflicts between competing agricultural, urban, and environmental demands for water. It needs to balance water project operations and allocations to support economic uses of water and restore environmental values held dear by the public. Without some resolution of these issues, the future of irrigated agriculture in California will be characterized by political and legal conflicts over the management and availability of both water and land resources.

Characteristics of Irrigation in California

Irrigation and water development in California are mutually dependent. Agricultural interests, allied with urban interests, traditionally have been a major influence in promoting federal and state water development projects in California and have been a large beneficiary of water development. California has more than 1,200 nonfederal and 181 federal dams and reservoirs, with a total storage capacity of approximately 42 million acre-feet. In addition, ground water makes up 22 percent of the total amount of water used in irrigation (33 percent in drought years). Agriculture uses 80 percent of the developed surface water supplies in California and 75 percent of the ground water used in the state (California Department of Water Resources, 1994).

California's farmers produce approximately 250 different crop and livestock commodities. Although specialty crops make up the backbone of the state's agricultural economy, dairy and cattle production and major field crops such as cotton and rice are important components of the state's market strength. Farming and irrigation practices vary among the distinct climatic and topographic regions of the state—from the rainy northern coast to the rice fields of the Sacramento Valley; to the corn crops in the Sacramento–San Joaquin delta islands; to the dry, highly productive San Joaquin Valley; to the coastal regions of Salinas Valley, Ventura, and San Diego; and to the hot, dry California desert. In addition to approximately 35 million acres of grazing land in California, there are five basic types of cropping systems used in the state. Irrigated field crops represent 60 percent of California's cropland; tree fruit, grapes, and nuts, about 18 percent; irrigated pasture, 11 percent; vegetable crops, 7 percent; and dryland crops, 4 percent. Many farms mix two or more of these systems and/or produce livestock or poultry (Demment et al., 1993). The irrigation technologies used with these systems are diverse, from gravity flood and furrow layouts to all piped, low-pressure, low-volume methods. During the 1987-1992 drought, farmers in the San Joaquin Valley and other regions became aware of the need to adopt more efficient technologies, many of them shortening furrows and adding return systems and others adopting drip irrigation systems.

The allocation and administration of surface water supplies are governed by a complex system of water rights and doctrines including the reasonable and beneficial use doctrine of California's Constitution, appropriative water rights and riparian water rights administered by the state, water rights (project) permits, the public trust doctrine, and an unregulated system of ground water rights. State and federal agencies play a major role in supplying and managing water for irrigation. The State Water Project delivers approximately 1.5 million acre-feet per year to farmers in the northern and southern parts of the Central Valley and on the central coast, and about 1 million acre-feet to urban users served by the Metropolitan Water District and other districts in Southern California. The federal Bureau of Reclamation delivers over 5 million acre-feet of water in an average year, about 20 percent of the total supply of irrigation water used in the state through the Central Valley Project. At the local level, irrigation districts, water districts, and other local agencies play a significant role in supplying water and drainage services to irrigators. In the 1950s and 1960s, irrigation and water districts built and enlarged dams to extend their service areas. Today, the role of districts extends well beyond supplying water to include drainage management, conservation and efficiency improvements, and ground water management.

Total applied water from all sources for agriculture in 1990 was 31 million acre-feet, of which total depletion was 24 million acre-feet. Twenty percent of the applied water was provided by ground water extraction (California Department of Water Resources, 1994).

Forces of Change and Responses

A three-way contest has been growing in California for the past two decades over the amounts of water used for urban and agricultural purposes and the amounts necessary to maintain environmental and ecological values in the state's rivers, delta, and wetlands. In a planning document, Bulletin 160-93 (California Department of Water Resources, 1994), the state estimates that urban water needs could grow from 8 million acre-feet in 1990 to 13 million acre-feet by the year 2020, and environmental water requirements could increase by 3 million acre-feet in order to protect endangered species and aquatic resources. The state also estimates that despite the outlook for greater transfers of water from agriculture to urban uses and loss of lands to urbanization and salinization, the total land and water dedicated to irrigation will decrease only about 5 percent. Other stakeholders, including environmental interests, contend that endangered species requirements, coupled with drainage problems, ground water overdraft, the possible loss of some water from the California Basin, and pressures of urban growth, will have a far more dramatic impact on irrigation. Farmers seem to be increasingly willing to transfer water and to make significant changes in farming technology in response to higher prices of water and other inputs. Irrigators and water districts, as well as urban users, have taken various steps over the past two decades to adjust to changing conditions of water supply and reliability in light of these factors, while negotiations, legislation, and litigation continue at the state, federal, and local levels to address the growing competition over developed water supplies. For instance, in 1994 agriculture, urban, and environmental interests, together with federal and state governments, agreed on a plan to meet water quality standards in the Bay-Delta region after a 20-year impasse.

The future of irrigation in California ultimately depends on how successfully irrigators and other water users can work together with state and federal regulators to resolve conflicts over water supplies and the degree to which irrigation's water and land use practices can be adapted to be more compatible with the needs and interests of other sectors of California's economy. Changes in water management policies by the Bureau of Reclamation and the recent state-federal agreements on operations to coordinate water quality management and flows for fish survival show that most parties are more willing to reach negotiated settlements than to live with the status quo.

Environmental Issues

The amount of additional water necessary to support environmental resources and water quality in California may be 3 million acre-feet or more. Whether or not these demands will mean a significant reduction in the supply of water available for irrigation depends on many factors. These include the uncertainty over the water that might be reallocated under the provisions of the Endangered Species Act, water quality laws, or the Public Trust doctrine. (The Public Trust doctrine embodies the principle that the state holds in trust title to tidelands and the beds of navigable waters for the beneficial use of the public and that public rights of access to and use of such areas are inalienable). Traditional public trust rights include navigation, commerce, and fishing. California law has expanded the traditional public trust uses to include protection of fish and wildlife, preservation of trust lands in their natural condition for scientific study, and scenic enjoyment and related open-space uses (California Department of Water Resources, 1994).

The assignment and administration of water rights and permits historically have been contentious, resulting in numerous legal and political battles over project operations and the construction of water facilities. The need to consider environmental water requirements has heightened the level of controversy among competing water users. In the 1984 decision regarding the way the State Water Resources Control Board set water quality standards in the delta, Appellate Court Judge Racanelli wrote that the state has broad authorities and obligations to enforce water quality objectives and water rights permits and conditions to protect the beneficial uses of the Sacramento–San Joaquin delta and San Francisco Bay (California Department of Water Resources, 1994). The State Water Resources Control Board is still in the process of determining how the water rights of all delta and upstream water users will be affected by the need to maintain downstream water quality standards.

The significance of endangered species protection in determining the availability of water for agriculture and urban uses became clear with the designation of winter run salmon and delta smelt as endangered species. Federal requirements for increased outflows out of the Sacramento River into San Francisco Bay and restrictions placed on water pumped from the delta have reduced the capability of the state and federal water projects to meet full delivery commitments to agricultural lands and urban water users. In a potentially important precedent, the state and federal agencies with jurisdiction over water resource management affecting the Sacramento–San Joaquin delta and San Francisco Bay (i.e., California Departments of Water Resources and Fish and Game, and the federal Bureau of Reclamation, Fish and Wildlife Service, Environmental Protection Agency, and National Marine Fisheries Service) joined together in December 1994 to develop a coordinated package of protections for the estuary. These include new EPA water quality standards, measures proposed by the National Marine Fisheries Service and the Fish and Wildlife Service to protect endangered species, and a plan for management of bureau (Central Valley Project) water dedicated to the environment. Together with urban, agricultural, and environmental interests, these agencies negotiated and agreed to provisions for operating the major projects and meeting fish protection and water quality standards for a period of the years during which a plan for the permanent operation and construction of any needed facilities will be identified (reference Principles for Agreement on Bay-Delta Standards Between the State of California and the Federal Government, December 15, 1994). While these legal and political processes continue, agricultural and urban water users may face water shortages and continued uncertainty over the availability and reliability of water supplies.

Another precedent for reallocating water supplies from consumptive uses to environmental uses was established with the Central Valley Project Improvement Act (P.L. 102-575), which dedicates 800,000 acre-feet annually to environmental protection and sets the goal of doubling anadromous fish populations by the year 2002. The act also imposes many operating criteria and allows transfers of water from agriculture to urban uses.

The impacts of these situations are already being felt in the San Joaquin Valley, but the actions are so recent that there is little documentation of reductions in irrigated area or indications of reductions that may take place in the near future. Further, California experienced a 6-year drought from 1986 to 1992, followed by one wet year in 1993, another dry year in 1994, and an extremely wet year in 1995. These variations make it difficult to separate the effects of the droughts from the actions to reallocate water from urban and agriculture to environmental purposes.

Other environmental constraints to irrigation include drainage-related water quality problems and the need for measures to manage and dispose of subsurface agricultural drainage. Approximately 2.5 million acres in the San Joaquin Valley face drainage and salinization problems. Without measures to manage and dispose of subsurface agricultural drainage, it may be necessary to cease irrigation on approximately 45,000 acres of land on the west side of the San Joaquin Valley by the year 2020. Farmers and water districts in the region have long pursued plans for completion of a concrete-lined aqueduct to collect and convey drainage waters to the Sacramento–San Joaquin delta for disposal. Environmental concerns over the water quality impacts of this project, as well as high costs, have stopped the project. Plans for expanding the use of drainage evaporation ponds also have met with strong opposition because the concentrations of trace elements such as selenium in the drain water are known to be harmful to waterfowl and other species. While many farmers are managing to control soil salinity and to continue production on these lands, the long-term outlook for continued irrigation of these lands is limited by the amount of salt that may be stored in the soil profile (National Research Council, 1989).

Farmers in the drainage-affected areas eventually may pursue water transfers and/or voluntary land retirement under a recently initiated federal-state land retirement program. Also, the state Department of Water Resources and the Bureau of Reclamation have experimented with drain water reclamation technologies. However, costs of recycling agricultural drain water are not competitive with other options for additional supplies at this time.

Water Supplies

The optimistic projections of the 1950s that irrigated acreage in California would continue to expand faded with the knowledge that water supplies are limited and the paucity of public support or capital to develop new projects. More and more, farmers and other water users in California recognize that any increase in water supplies is more likely to come through improved management of existing supplies than through the construction of new water supply projects. Demands for water in urban and environmental uses, coupled with demands for irrigation, exceed available supplies. In the face of these limited supplies, the total irrigated acreage in California is expected to remain at or below the current level of 9.2 million acres, decreasing to approximately 8.8 million acres by the year 2020 (California Department of Water Resources, 1994). The qualitative impacts of limited water supplies are likely to be more significant than the net reductions in irrigated acreage as farmers contend with continuing ground water overdraft and rising costs.

In the past, ground water overdraft has been managed, but not always eliminated, with the importation of surface water. Many urban areas have successfully controlled ground water overdraft only to face more difficult problems of chemical contamination of their ground water basins. In the agricultural Central Valley and in northern California, where ground water basins are not adjudicated or managed with imported water supplies, water storage will be depleted and pumping depths will increase, making ground water pumping uneconomical compared with other sources of water. Ground water pumping increased dramatically during the 1987-1992 drought in the San Joaquin Valley and other regions where irrigators faced water supplies 15 to 50 percent lower than historic project deliveries. The quality of ground water in some basins, particularly those in the central coast region, may deteriorate as a result of salt water intrusion. These localized ground water problems may lead to some decrease in the amount of land that can be irrigated in a particular region.

Largely in response to recent droughts, urban and agricultural water districts are working on plans to formalize and institutionalize water conservation as a standard element of water management. In regions where the cost of water is high and/or the reliability of water supplies is uncertain, farmers have improved their irrigation methods and equipment in recent years and are able to achieve high irrigation efficiencies. Water districts in the central and southern San Joaquin Valley have water application efficiencies generally higher than 70 percent. With the support of the state, agricultural districts have worked to develop a program for voluntary "efficient water management practices." However, few agricultural representatives believe that conservation measures will yield large water savings, especially in cases where surface water is used in conjunction with ground water from an overdrafted basin.

In urban areas, landscape and turf irrigation is a significant factor in water consumption, using some 6 million acre-feet each year in California. Conservation practices and reclamation account for a growing portion of the water used in urban irrigation, especially on golf courses. San Francisco Bay area water suppliers, led by the city of Santa Clara, have proposed a project to transport treated wastewater to the San Joaquin Valley for agricultural use in lieu of meeting more costly treatment and disposal EPA requirements for ocean disposal of treated wastewater.

Water transfers have long been used by water districts as a way to address short-term surpluses and shortages from one farmer to another. In the past 20 years the interconnection of water delivery systems, particularly those of the large state and federal water projects, has provided even greater opportunities for exchanging water among users. Increasingly, urban water suppliers are looking to transfers from agricultural areas as an important source of water.

Although only relatively small amounts of water have been transferred to date from agricultural to urban users, the number of transactions is growing, and various types of water transfer arrangements are emerging. These include the purchase of agricultural land and agricultural water contracts, installation of water conservation works in exchange for rights to the salvaged water, and leases or options for a limited number of years (e.g., 7 out of 15 years) that still allow the agricultural seller to continue to operate part of the time. Exchanges can also be made by the delivery of water into ground water basins. In 1995, in an agreement between agricultural and urban users in the State Water Project, about 10 percent (130,000 acre-feet) of the agricultural water was earmarked for transfer to urban users from agricultural districts where the cost of water had become too high for profitable farming. In exchange, modifications are being made to the rules governing allocations of shortages during water-deficient years.

One long-standing impediment to the exchange or transfer of federally managed water was the restriction that precluded transfers of Central Valley Project water to users outside the project boundary. This constraint effectively has been eliminated by the Central Valley Project Improvement Act (P.L. 102-575). For most of the past decade, the California legislature has been debating legislative solutions for water transfers to remove some of the uncertainties from both buyers and sellers in regard to preservation of water rights and to relieve or limit third-party effects.

Examples of transfer arrangements in California include the 1992 purchase of water from the Devils Den Water District located in Kern County by Castaic Lake Water District (an urban district located northwest of Los Angeles). Castaic Lake Water District will continue to lease out some of the farmland for operation as economic conditions and water needs allow, but will move water into its area as urban demands increase. A transfer of 3,500 acre-feet of State Water Project water is being made from Berrenda Mesa Water District in Kern County to the city of San Ramon, near Oakland, under the 1995 agreement referred to above. In 1994 a farmer in the Central Valley Project service area proposed to transfer 32,000 acre-feet of water over 15 years to the Metropolitan Water District of Southern California (MWD) for approximately $5.6 million. The Metropolitan Water District would be able to purchase about 4,600 acre-feet in each of any 7 years of its choice.

At present, no estimates have been made of the total amount of water that may eventually be transferred from agricultural to urban use. However, California's population is projected to increase from 30 million in 1990 to 49 million in 2020, or 63 percent. Even assuming extensive water conservation and the implementation of a number of water supply improvements, including wastewater reclamation, the state anticipates that demands for water will exceed developed supplies by 2 to 4 million acre-feet per year (California Department of Water Resources, 1994). This situation will tend to drive up the value of water, increase the prices that urban areas are willing to pay and increase the pressure on agriculture to transfer water. Increased demands for water transfers also would provide political momentum for addressing remaining obstacles to water transfers, including concerns over third-party effects and environmental restrictions on long-distance transfers across the delta.

Approximately 380,000 acre-feet of municipal wastewater is reclaimed and recycled in California to augment urban supplies. This is up from 270,000 acre-feet in 1987. Most of this water is used for ground water recharge, agricultural irrigation, and landscape irrigation (California Department of Water Resources, 1994). Wastewater (recycled water) is a valuable commodity, and its use to irrigate golf courses has risen dramatically in the past several years. It also may be used to restore and maintain wetlands located on large turfgrass sites. There are approximately 1,000 golf courses in California, each with approximately 125 acres of green turfgrass (Steinbergs, 1994). Recycled water, especially in time of drought, can be used to keep part or all of a golf course green and playable.

Urban Expansion

Irrigated agriculture also is feeling the effects of urban expansion as agricultural land is converted to urban and suburban uses. The California Department of Conservation estimates that 32,000 acres were converted from prime irrigated farmland to urban uses from 1984 to 1990. The Department of Water Resources, in looking ahead to future water needs, estimates that 300,000 acres of irrigated farmland may be converted to urban uses between 1990 and 2020 (California Department of Water Resources, 1994). The impact would be felt throughout the Central Valley and Southern California. This trend principally affects the acreage that grows high-value crops such as stonefruits, citrus, nuts, vegetables, and dairy products, which in turn are being moved to new locations. There are efforts in some areas to establish policies to preserve agricultural land in and near urban locations, and so far the loss of good irrigable lands is not a serious problem because there has always been an ample supply of other irrigable land that can take over any available production niche. While many citizens support preservation of agricultural lands for its economic and aesthetic value, few local governments are willing to place strong limitations on the conversion of farmland to urban developments.

Rising Costs

Some of the most serious challenges to the availability of water for agriculture, and to the future of irrigation, come from increases in the cost of water supplied by both the state and the federal water projects. The cost of water from the State Water Project has increased to three times the cost projected when the project water was first delivered in 1970, that is, from $25 to $75. In areas where project water also must be pumped up in order to reach individual farms, the cost of water, influenced by today's high cost of electric energy, may make it unprofitable to grow lower-value crops such as cotton. Increases in water pricing through legislative and policy reforms, such as the Central Valley Project Improvement Act, also may result in changes in cropping patterns and irrigation practices at the local level.

Conclusion

With increasing competition over water supplies, a likely increase in water transfer activity, rising water prices, and rising land prices and input costs, it appears that California's irrigators will have to continue to adjust and adapt practices to remain competitive. Over the next 25 years, it is unlikely that restrictions on the availability of good-quality irrigable lands in satisfactory climatic zones or on the availability of water will jeopardize California's competitiveness in markets for specialty crops. Major field crops, such as cotton and rice, and the dairy and cattle industries also are likely to remain competitive because of their consistently high quality. However, rising costs of farm production inputs, including increased costs associated with changes in water management, are a potential impediment to California's ability to compete in foreign markets. Few agricultural experts predict that California agriculture will lose significant shares of the markets it now serves. However, California farmers face a period of great readjustment in the years ahead as the conditions for water availability, water price, and the costs of farm inputs change.

The primary constraints to irrigation's water supply come from increasing environmental and urban water demands. By 2020, urban demands are projected to increase 62 percent over current levels. Much of this increase is associated with demands for water for landscaping and outside uses, which range from 30 percent in coastal climates to 60 percent in hot inland areas. Urban water suppliers are spending large amounts on conservation, reclamation, water transfers, and conjunctive use to offset the need for additional new water supplies to meet urban demands.

Responses to the challenges of limited water supplies are evident within the agricultural community as well as in water management institutions and agencies. With uncertainties over environmental requirements and water supplies likely to continue, local institutions in California, particularly water districts, can be expected to assume greater responsibilities for water management and project operations. At the farm and district levels, the agricultural community has responded to growing uncertainty over water supplies and other pressures by undertaking improved irrigation practices, increased water transfers, conjunctive use, technological improvements, and management of water in drainage-affected areas. Districts are adding water conservation specialists to their staffs, making water conservation plans, and using price structure to influence water use. Water supply constraints on irrigators will change how farmers do things such as decide on cropping patterns or relocation of some specialty crop production or improve irrigation efficiency. With the total irrigated area shrinking, and with water becoming more costly and more subject to reallocation, California farmers face many dfficulties to maintain their relative competitiveness in local, domestic, and international markets.

Questions about the quantity and reliability of the irrigation water supply are manifest in numerous negotiations, legislative proposals, and legal battles in federal, state, and local venues. Environmental demands will continue to be the major factor behind continuing negotiations and proceedings to reallocate developed water supplies.

Concern about drainage-related pollution problems may result in selective land retirement and increased water marketing from drainage-impacted areas. The Bureau of Reclamation and the state have developed a program for voluntary retirement of drainage-impacted lands to yield water quality, habitat, and water supply benefits. Farmers and districts also may negotiate short-term or multiple-year transfers to manage shallow ground water in poorly drained areas. Water managers and resource agencies also will look to technological innovations, including reverse osmosis and reclamation, to address the drainage problem.

The most serious threat to the overall competitiveness of irrigation in California compared to the rest of the nation has been the uncertainty and prolonged controversy over the allocation of developed water supplies among agricultural, urban, and environmental uses. Conflicts over water management were seen as a threat to the overall investment climate during the recent economic downturn in California. The December 1994 agreement among state and federal water resources and environment management agencies that provides a 3-year period of stability in bay-delta water operations while a longer-term plan is developed marks a new era of willingness of all parties to reach a satisfactory solution. The solution will include both water operation provisions and physical works for better water management. Although it may be difficult for all interest groups to work within this framework for agreement, most agree that an extraordinary effort for cooperation and conflict resolution is critical to ensure a resulting operation plan that will provide for water management in California to satisfy all the future agricultural, urban, and environmental uses.

IRRIGATION IN THE PACIFIC NORTHWEST: ENVIRONMENTAL DEMANDS, TRIBAL TREATY RIGHTS, AND INSTITUTIONAL CHANGE

By far the most significant issue confronting irrigation in the Pacific Northwest is competition for water among agriculture, environmental, tribal, and, to a more limited extent, urban uses. The largest single competing demand is for water to help restore endangered salmon populations. The Columbia River basin provides ample evidence of the difficulties and transformations being experienced in the Pacific Northwest by the various water-related interests, including the irrigation community. Also, it is a basin that calls for a new paradigm that advances environmental restoration, ecosystem management, protection of tribal treaty rights, and collaborative decision making.

The Pacific Northwest region—Idaho, Montana, Oregon, and Washington—is the leading producer of many crops in the United States, including apples, hops, mint, potatoes, and cherries. Irrigated agriculture also supports extensive related industries and infrastructure, such as processing, packaging, shipping, and transportation. Irrigation in the region began on a small scale, started by several American Indian tribes at mission sites near Walla Walla and Yakima, Washington, and Lewiston, Idaho. With population growth driven by trading and mining, irrigated acreage increased to one-half million acres by 1900. Large-scale irrigation projects authorized by passage of the Desert Land Act of 1877, the Dawes Allotment Act of 1892, the Carey Act of 1894, and, in particular, the Reclamation Act of 1902 resulted in substantial irrigation development. Today, more than 50 percent of the irrigated land in the region receives water from reclamation projects. In addition, power generated at such dams provides the low-cost energy required to pump irrigation water.

The Columbia River basin, which includes the Snake River, covers over two-thirds of the four Pacific Northwest states. The Columbia River is controlled with a vast and complex combination of federal and nonfederal facilities. The Corps of Engineers and Bureau of Reclamation have built major dams on the river and its tributaries. The dams support flood control, irrigation, navigation, hydroelectric power generation, recreation, fish, wildlife, and water quality. The responsibility for managing these uses is shared by a number of federal, state, tribal, and local agencies.

There are 14 federally recognized tribes in the Columbia River basin. Because of the federal government's Indian trust responsibilities and the government-to-government relationship with tribes, special efforts are being taken to provide for meaningful participation in coordination with tribal governments in various approaches needed to respond to the changes, especially in fisheries management and recovery.

Characteristics of Irrigation in the Northwest

In 1990 the Pacific Northwest had a total of 10 million acres of irrigated lands. Over 7 million acres are irrigated in the Columbia River basin alone. Idaho has the largest irrigated acreage in the basin, 3.4 million acres (48.5 percent of the region's total irrigated acreage), while Washington has 1.9 million acres (27 percent), and Oregon 1.3 million acres (18.5 percent), with Montana at 0.4 million acres (6 percent) (Moore et al., 1987).

In Oregon and Idaho, agriculture is the leading industry. In Oregon the industry represents 17 percent of the economy, 25 percent if considering value-added services. In Idaho, production agriculture and food processing account for over one-third (35.9 percent) of the gross state product (Idaho Agricultural Statistics Service, 1992).

Irrigation efficiencies vary significantly. Approximately 90 percent of total water withdrawn in the region is for irrigation. Surface water irrigation accounts for 75 percent and ground water 28 percent. Irrigation directly out of the Columbia Basin in the four states is significant when considering the average seasonal water flow of the basin. The net irrigation depletion, essentially diversion minus return flows, is estimated at 14 percent of the average seasonal flow for the Columbia River basin. It is estimated that over 43 percent of irrigation in the region is with gravity systems and 57 percent with sprinkler systems. Conversion to sprinkler systems is occurring rapidly, specifically on lands with high-value crops and in areas subject to water shortages (Bonneville Power Administration, 1993).

Irrigation in the region is characterized by a high degree of diversification and intensive land uses, varying from lands under hay production to lands with intensive irrigation of fruits and vegetables. The value of crop production ranges from $150 per acre for hay, using flood irrigation, to $6,000 per acre for apples under drip irrigation. In 1991 the gross crop value per acre in Montana was $216, Idaho $457, Oregon $578, and Washington $1,400. The value of crops per acre from Washington is third in the nation, after California and Arizona.

The size of the farms varies from 40 acres to large operations of 10,000 to 20,000 acres. Some lands that were once classified nonirrigable under gravity are now irrigated with center pivot systems.

The region contains approximately 33 million acres of land potentially irrigable with favorable soils and climate. However, many of these lands have little or no prospect for irrigation because of limited water availability or other considerations (e.g., markets, environmental restrictions, and concerns over fisheries resources). It is estimated that 20 percent of the irrigated area is subject to water shortages. Most streams in the region are fully appropriated or overappropriated, especially in adjudicated basins or where adjudications are underway (e.g., the Yakima and Snake rivers). Large-scale irrigation proposals are very controversial. One example is the Bureau of Reclamation proposal to add over 87,000 acres to the existing Columbia Basin Irrigation Project in Washington state, served by pumping water from behind Grand Coulee Dam. The opposition stems from the argument that it is unwise to continue diversions from the river before devising a long-term strategy to address the salmon recovery and the conflicts among existing competing uses.

The four states of the Pacific Northwest have similar water codes and laws, with the appropriation doctrine as the framework for water allocation. A number of major adjudications are underway in Idaho, Washington, and Montana, including several negotiations between states, tribes, and the federal government. All four states have active irrigation organizations. Most of these organizations are political subdivisions of state and local governments and municipal corporations. In Idaho and Washington, 74 percent of the irrigated land is under the control of irrigation districts, 56 percent in Oregon, and 80 percent in Montana. These districts have authority to provide water for lands within the district boundaries, to acquire facilities, to enter into agreements, and to use state loans and grants for conservation, water development, fish passage, and dry leases.

With increasing competition over water supplies, all four states have enacted policies for water conservation and protection of instream flows. Water banking, water leasing, cost-sharing programs, and other incentives are being promoted in the region to facilitate conservation, efficiency, and reallocation of water for other public benefits. Oregon and Washington have adopted statutes that allow the conserved portion of a senior water right to be transferred or leased to other users. This change offers promise both to the irrigation community and to the public through potential reallocation of "saved" water without loss of priority.

Forces of Change and Responses

The Columbia River has been described as a "system under stress" and poses major challenges on how better to resolve the conflicts among competing uses and resources and how to involve the large number of interests in the resolution process. In the 1930s and 1940s, hydropower development on the Columbia River system was crucial to aluminum processing and the military victory in World War II. Continued growth in the region has put steadily increasing pressure on the river system. The prevailing perception in the region is that there is no longer enough water to fully satisfy all of the demands. Intense competition and bitterly contested decisions over water are common. These circumstances, and in particular the Endangered Species Act listings and American Indian treaty rights, have resulted in ill-defined institutional arrangements and an unpredictable and unduly complex decision-making process.

At the present time, uncertainty is the most significant characteristic throughout the region. Short-term and long-term decisions are difficult to make given the current level of uncertainty.

Competing Uses, Environmental Concerns, and Treaty Rights

In addition to irrigation, the Columbia and Snake rivers have been heavily developed for hydroelectric power. The system includes 30 federal hydroelectric projects plus several local and private projects. Fifteen Northwest utilities, the Corps of Engineers, and the Bureau of Reclamation are the primary beneficiaries of the dams.

Hydroelectric dams on the Columbia and Snake rivers are the foundation of the Pacific Northwest's power supply. Hydroelectric power is a critical factor in the thriving economy of the region. Over two-thirds of the region's power is from hydroelectric generation. Power produced at dams in the region serves customers both locally and in other states. Hydropower generated at Reclamation dams provides low-cost power needed to pump water and operate sprinkler systems.

Over the years, however, hydropower development has altered, degraded, and eliminated fish habitat in the Columbia and Snake rivers. The hydropower system is undergoing operational changes as a result of efforts to facilitate salmon recovery and has run into conflict with American Indian treaty rights. Drafting the reservoirs at certain times of the year to provide additional flows for fish passage will decrease generating efficiency and may result in wholesale rate increases.

Box 5.1
Cooperative Planning in the Klamath River Basin

The Klamath River basin is a 1,000-square-mile basin in Oregon and California encompassing a mix of federal, state, tribal, and local water users dependent on the river. The Bureau of Reclamation built the Klamath Project in 1940 and currently services 125,000 acres of irrigated land under contracts signed with water users. The project includes a series of impoundments on the Klamath and Lost rivers, four major diversion structures, several tens of miles of lateral and pumping facilities. Four Indian tribes, including the Klamath, Hupa, Yurok, and Karuk, have extensive treaty rights to fish in the upper reaches of the Klamath, currently impounded by the Bureau facility, as well as in the lower reaches of the river in northern California. The conflict over water supplies and distribution has been heightened in recent years by prolonged drought, and it pits contract holders organized under state law against the interests of tribal treaty fishing rights, which predate contract holders. The response of this community and the federal agencies to this conflict has spawned the development of new institutional arrangements that are providing concrete solutions to water management problems in the Klamath River basin.

Under a recent imperative to resolve the conflicts over water use and management in the basin, the Bureau is in the process of developing an operating plan for the Klamath Project. As part of this process, the four tribes, local water users, the Fish and Wildlife Service, the Bureau of Reclamation, and the Klamath River Restoration Task Force have embarked on an effort to provide for cooperative river basin planning in the context of existing water rights and governmental authorities. In this case, the Bureau has trust obligations to protect the tribes' treaty rights, but it must also honor its contract commitments to the non-Indian irrigation community in the context of state law.

The tribes, in cooperation with the Department of the Interior, have created a template for government-to-government relations, which guides the nature, protocol, and activities involved in ensuring that tribal trust obligations are incorporated in the plan. The tribes participate in technical subcommittee meetings with other water users. The Bureau conducts regular working group meetings consisting of the local water users, tribal government representatives, and other federal interests. Additional consultation is conducted with the tribal elected chairs and federal officials.

The parties in the Klamath River basin are building a long-term institutional structure for cooperative water management, based on a system of laws, historical practice, and the application of science and technology to the resolution of disputes involving the use of water for irrigation, fisheries, and wildlife purposes.

The reservoirs behind dams also created a navigation industry, and inland waterway navigation and water transportation are key elements of the Pacific Northwest's regional development and growth. Economic benefits occur to cities, communities, and the region. Agricultural producers rely on the inexpensive, reliable, and accessible water transportation system. But the higher river flows needed to improve salmon survival may lower the reservoirs and hinder transportation.

Recreation also competes for water. The recreational facilities of the Columbia River basin attract visitors from all parts of the country. Boating, windsurfing, and fishing are a few of the popular recreational activities throughout the basin, especially in the summer. They provide major economic benefit to various communities. Recreation relies on stable flows and water levels. Changing river flows and reservoirs levels will affect the quality of the recreational activities.

Another competing demand is municipal and industrial water supply. The combined population of Oregon, Washington, Idaho, and Montana in 1990 was over 9.5 million people, increasing at an annual growth rate of 2.2 percent. The population of the Pacific Northwest is projected to grow by about 30 percent between 1990 and 2010. The majority of the growth is occurring in urban areas. For example, the Willamette Valley area is the highest agricultural producer in Oregon and is also home to 70 percent of the state's total population; major urban development is encroaching along the valley corridor and competing with irrigation for the limited water resources.

Flood control is an important priority for river operations. Construction and operation of projects in the United States and Canada have dramatically reduced the damage caused by floods from the Columbia River. Drawdowns and releases to mitigate floods are being evaluated to maximize the availability of water for fish, especially in low-water years.

The water quality in the Columbia basin is relatively good. There are few sources of industrial and municipal waste. However, nonpoint sources, irrigation withdrawals and returns, dams, and water releases are major issues in many tributaries and reservoirs. Irrigation return flows are a major source of nonpoint-source pollution. Studies conducted along the Yakima, Umatilla, and Okanogan rivers identified water quality degradation, especially during summer months, from both irrigation withdrawals and return flows carrying nutrients, pesticides, and sediments. Water diversions for irrigation reduce flow rates, causing loss of spawning areas, higher temperature, and higher concentration of pollutants. Also, unscreened or improperly screened diversions result in fish mortality.

Anadromous fish of the Columbia and Snake river basins are in the center of much economic, political, and social debate regarding the lengths to which the region should go to restore the salmon and steelhead. Of 320 stocks, 261 are (or were) found in the waters of Oregon, Idaho, and Washington. Of this total, 106 major stocks of native salmon and steelhead are now extinct, and another 102 are at high risk of extinction (American Fisheries Society, 1991). The Northwest Power Planning Council, a regional organization, estimates that Columbia River basin salmon and steelhead runs ranged between 10 and 16 million wild fish prior to modern development. Since 1970, the number of adult salmon and steelhead entering the Columbia ranged from 0.9 million in 1983, to 2.9 million in 1986, to 1.1 million in 1990 (Northwest Power Planning Council, 1994).

The decline of wild runs has been so severe that three stocks are now listed under the Endangered Species Act. The threats affecting each stock include habitat destruction and alteration, overfishing, disease, and competition with hatchery fish. Agricultural activities also have harmed the salmon habitat throughout the basin. The adverse effects of irrigation include increased temperatures; increased erosion and sedimentation; reduced flows in spawning areas, especially in tributaries; blockage of fish migration; loss due to unscreened diversions; and degradation of water quality. To counter these negative influences, salmon recovery efforts are going forward, including studies of new reservoir sites, new reservoir drawdown strategies, fish hatchery operational changes, and water management opportunities such as water conservation, transfer, marketing, pricing, and conjunctive use of ground and surface water.

Salmon are more than a source of food in the Pacific Northwest—they have a spiritual role in Northwest American Indian culture. American Indians have strong ties to the river system and the salmon. Northwest tribes hold significant claims for water for irrigation and protection of fisheries and other purposes. As a result of treaties and federal statutes, Northwest tribes hold and exercise rights beyond their reservation boundaries. Conflicts between various tribes and states over tribal fisheries have resulted in the court reaffirming the tribal rights to share equitably in the harvest of the salmon and steelhead and to continue fishing in their usual and accustomed places. The tribes now participate in decisions affecting management and recovery of fisheries resources. Many tribes feel, however, that more emphasis is being placed on the cost of environmental protection and less on its value.

Although the focus in the region is on salmon and steelhead, wildlife and resident fish, which live year-round in the rivers and reservoirs are also very important. Several national wildlife refuges are located adjacent to the Columbia and Snake rivers. Many of the riparian areas and wetlands were established as a result of the construction of the dams and reservoirs. They rely on stable or minimum flows and reservoir level fluctuations. Resident fish have been impacted by the reservoirs, diversions, and habitat destruction. Recovery and preservation of resident fish are being addressed by the various federal, state, tribal, and regional institutions. Watershed management and cooperative habitat protection are being promoted and implemented in several areas.

Regional Institutional Processes and Responses

The Columbia River system is complex, and changes to benefit any particular use will inevitably affect other uses. As the pressures in the region continue to grow because of urbanization, economic development, and environmental and resource management concerns, the competition for control of water resources will become even more intense. Demand for water for environmental uses, particularly the protection of endangered salmon species, is the most controversial issue affecting irrigators in the Pacific Northwest.

Agencies have numerous processes and responses underway in the region, ranging from conservation studies and water acquisition to salmon recovery and rebuilding plans. The Northwest Pacific is known for its megabureaucracy. Indeed, the challenge in the region revolves around integrating and coordinating the activities of numerous agencies at the state, tribal, local, and federal levels. The Corps of Engineers and Bureau of Reclamation built and operate dams for flood control, irrigation, and hydropower. The Bonneville Power Administration sells the power from federal dams to utilities throughout the region. Although these three agencies have primary responsibilities for the river operation, other federal, regional, tribal, state, and local governments play critical roles. Historically, each of these agencies had primary mandates. However, that is changing, and new responsibilities and missions have been added, such as fisheries, recreation, and tribal trust. Conflicts between and within the agencies regarding river operations are common.

Several biological, hydrological, and engineering studies are underway to review river operations. For example, the Bureau of Reclamation is conducting a review of operations in the Snake River basin with the goal to develop a model to better manage water and related resources at 10 Reclamation projects. Some of these projects provide extensive irrigation water. Changes in reservoir operation, storage, and diversions are likely to be recommended.

In addition, the Bonneville Power Administration, Corps of Engineers, and Bureau of Reclamation are jointly conducting a review of the operation of the 14 Columbia River system hydroprojects (Bonneville Power Administration et al., 1994). The goal of the "System Operation Review" is to develop a system operating strategy and a regional forum for allowing interested parties, beyond these federal agencies, a long-term role in system planning. No preferred operating strategy has been selected.

The National Marine Fisheries Service is also studying threatened and endangered salmon, and it is responsible for developing a recovery plan and issuing biological opinions. The Fish and Wildlife Service has similar responsibility for other species. The most recent biological opinion recommended only minor changes in the existing system, to the great concern of the Northwest tribes and others (Columbia River Intertribal Fish Commission, letter to William Stelle, Regional Director for National Marine Fisheries Service, February 10, 1995).

In the Northwest, institutions are trying to settle the problem of the salmon using a mix of local and regional activities. For example, when the Endangered Species Act petitions were filed, political pressure mounted for the region to find its own solutions before the actual listing. A "Salmon Summit" was convened to develop and put into place a plan to rebuild the stocks targeted by the Endangered Species Act petition. The summit resulted in a plan in 1991 that called for various measures, including providing additional storage water, installing screens and bypass facilities at irrigation diversions, testing the effects of drawdown, and maximizing fish transport.

The Northwest Power Planning Council a regional organization made up of representatives of the four states, develops Power Plans and Fish and Wildlife Programs for the Columbia River basin. In 1991 the council, in response to potential listing of endangered species and in response to a request of the governors of the four states following the Salmon Summit, embarked on a salmon rebuilding program. The position of the council is that those who use the river should bear their share of the costs of measures needed to rebuild fish stocks. Although the council does not play a direct role in shaping the future of irrigation, it has incorporated in its Fish and Wildlife Program actions that will have an impact on agricultural irrigation.

The NPPC has approved several actions intended to assist in the recovery of the Snake River salmon runs. These actions, which are being implemented by the Bureau of Reclamation and the states, include limited future water withdrawals, flow augmentation, water acquisition, new storage assessment, and uncontracted storage space. These significant actions affect and involve the irrigation community in all four states, but especially in Idaho, Oregon, and Washington. For example, in the Snake River basin a report prepared for the Northwest Power Planning Council and the Bonneville Power Administration has identified water management opportunities in Oregon and Idaho to secure at least one million acre-feet of water per year for the Snake River basin. The findings and conclusions of the study show that by using water use efficiencies, market mechanisms, water transactions, and land fallowing and implementing on-farm management and conservation measures, at least one million acre-feet of water can be acquired annually from existing uses, although no water acquisition has occurred yet.

At the individual state level, Oregon and Washington embarked on an effort to restrict water withdrawals from the Columbia and Snake rivers and their tributaries, following the listing of the Snake River sockeye. This restrictive policy, coupled with an aggressive instream flow program, places most agricultural water users in the position of having to become more efficient with their existing water use.

At the local level, watershed management and regional planning programs involving irrigation districts and individuals are working to improve water quality and quantity and to identify and carry out irrigation water management improvements on the ground. The states of Washington and Oregon have provided grants and loans to help these efforts.

In addition, irrigation and hydropower users and environmental and tribal representatives are participating in local efforts to design solutions to water management problems. Incentive-based conservation programs are being implemented throughout the region to encourage conservation, reallocation, and water acquisition. In Oregon a new organization, "The Oregon Water Trust," patterned after the Nature Conservancy, was formed for the purpose of purchasing water mostly from irrigators for instream uses. The irrigation community is playing an important role in defining and implementing the trust. In the Deschutes Basin, the Environmental Defense Fund, the Confederated Tribes of the Warm Springs Reservation of Oregon, and the irrigation districts have entered into a contract with the Bureau of Reclamation for a demonstration project to implement water conservation and secure conserved water and to review the institutional constraints and propose changes to make water leasing projects more effective.

The Umatilla example is another illustration that fish enhancement and irrigation can be compatible. The Umatilla River is a tributary of the Columbia River that drains farmland and parts of the Umatilla Indian Reservation in northeast Oregon. Irrigation diversions had dried up the river for over 20 years, and its salmon runs were history. Broad political support was built for a comprehensive restoration project. The Umatilla Tribes, Oregon fish and wildlife agencies, and the irrigators have restored some fish runs and are working to restore flows to the lower river and keep the farm economy whole. The irrigation water now taken from the Umatilla will be replaced by water pumped from the Lower Columbia River.

Although there are cooperative efforts underway to recover fish populations, and some local successes, the enormous scope of the salmon recovery effort, traditional water management policies and politics, the inadequacy of the existing institutions, and the multitude of competing interests are major constraints.

Conclusion

The future of irrigation in the Pacific Northwest is closely related to the future of the Columbia River. The decision to recover salmon in the Northwest involves trade-offs and will require broad cooperation. Opportunities and tools exist to address the needs of the salmon and steelhead but not without costs. How significantly agricultural irrigation will be affected is going to depend on its willingness to participate and contribute to the enormous effort of rebuilding the salmon populations.

Although today there is no consensus on how the conflicts and changes should be resolved, there is more of an awareness of the limits that individual state, tribal, and federal governments have in resolving these highly complex and controversial conflicts. It is obvious that accommodation of the many demands cannot be done without using a comprehensive ecosystem approach and unprecedented legal and institutional collaboration among the multiusers, multiinterests, and multijurisdictions.

Like the Columbia River itself, the challenge of providing water and other measures to protect salmon binds together all water users in the basin. In this context, it is stakeholders who develop more effective means to resolve conflicts, develop consensus, provide flexibility to respond to changing needs, improve the efficiency of water for irrigation, and optimize the allocation of water resources. Full public participation must be sought, and economic and social impacts must be considered.

IRRIGATED AGRICULTURE IN FLORIDA: INSTITUTIONS AND INDUSTRY IN TRANSITION

Unlike most areas dependent on irrigation, annual rainfall in Florida consistently exceeds evapotranspiration. Nevertheless, irrigation is required by the marked seasonality of rainfall in Florida. The ability to apply supplemental water during the dry spring months is essential to produce agricultural crops and to maintain urban landscapes.

As with many states in the arid West, the competition for water, expanding environmental constraints, and rapidly changing market conditions are major factors influencing irrigation in the Southeast. In addition, differences in climate, natural environment, soils, and prevailing cropping patterns create distinct challenges for the management of irrigation systems in Florida (Camp et al., 1990).

Irrigation in Florida provides a hedge against droughts and freezes, and it is an important element in achieving optimal yields. Reliable irrigation allows farmers to produce high-value crops and to meet market windows that are closed to other parts of the country because of climate. Reliance on ground water is the rule for the majority of Florida agriculture. In spite of high annual rainfall, surface supplies are the primary irrigation source only in the region adjacent to Lake Okeechobee, which includes Florida's sugarcane acreage and important amounts of vegetables, citrus, and sod.

Characteristics of Irrigation in Florida

In 1950, irrigated cropland in Florida was estimated at 300,000 acres. Following the droughts of the early 1960s, irrigated acreage jumped to over one million acres. By 1978 the irrigated area had climbed to over 2 million acres, only to drop by 400,000 acres because of freezes in the 1980s. Withdrawals totaled 3.8 million gallons per day, of which 53 percent was ground water and 47 percent was surface water. Agricultural expansion over the next decade raised the irrigated area to 2.1 million acres by 1992.

Agriculture was the largest user of water in Florida in 1990 (Marella, 1992). Citrus crops account for the largest acreage and withdrawals for irrigation (33 percent). Other crops with significant water use are sugarcane (22 percent), sod (5 percent), and turfgrass on golf courses (5 percent) (Marella, 1992). The *1987 Census of Agriculture* ranked Florida fourth nationally in market value of agricultural products sold from irrigated farms ($3.3 billion). In 1990, Florida had the largest irrigation withdrawals of any state east of the Mississippi River (Marella, 1992). Florida applies more irrigation water per acre than does Texas, even though rainfall in Florida far exceeds that in Texas (Bajwa, 1985).

Since 1972, Florida has been governed by one of the most progressive water resource management statutes in the country. In response to one of the worst droughts in the state's history in 1971, and public concern about the need for oversight of water resource management in the face of rapid population growth, the state created five regional Water Management Districts (WMDs). These agencies have the legal authority and financial capacity to manage water comprehensively—through regulation of all water use and surface water management, setting criteria for water quality and wetland protection, and imposing conservation and water shortage management. They also have evolved into the largest landowners in the state through well-funded land acquisition programs designed to preserve Florida's environmental and water-related resources.

Water in Florida belongs to the people collectively and can only be used according to administrative and resource protection criteria set by the WMDs. The license to use water is a temporary benefit that is reevaluated every 5 to 10 years. This exposes large irrigation users to possible reallocation to other consumptive uses, such as potable supply for cities. It also provides flexibility for changing social and political values, such as wetland protection, and allows the WMDs to mandate the adoption of the most efficient irrigation technology where that is warranted.

The WMDs began as water resource agencies dedicated to water supply and flood control. They have evolved into powerful and well-financed entities dominated by environmental protection and land acquisition and management mandates in addition to their traditional water resource roles.

The comprehensive legal framework enacted in 1972 has allowed the WMDs to preside over an orderly allocation process as the state's abundant water was made available to fuel agricultural and urban growth. Now they are facing the prospect of having to tell some potential users "no," and even some existing users "no more." This process will not be nearly as orderly as the initial exercise of their authority. The institution itself is under pressure to a degree it has never been in the past. It is too early to tell which issue will dominate in the next evolutionary phase of water management in Florida, but water supply is clearly the issue that will focus the spotlight on the Water Management Districts.

Forces of Change and Responses

Despite averaging over 50 inches of rain per year, Florida is facing challenges to the use of water for irrigation that are strikingly similar to those in California, namely, growing environmental and urban demands for water. The urban population is growing steadily and is finding its traditional water sources no longer sufficient. Florida's population has doubled in the past 20 years and is slated to reach 16 million by the year 2000. This growth is an unrelenting challenge to water management that is testing the state's institutional capacity to balance the competing demands on the natural resources. In addition, the people of Florida are beginning to question some of the environmental trade-offs that past generations were willing to make to encourage economic development. In a state dominated by urban population centers, the lack of understanding and acceptance of the value of agriculture poses a constant challenge to the irrigation industry.

Environmental Issues

The extraordinary commitment that Florida has made to irrigated agriculture has resulted in significant impacts to water-related environmental resources. Water levels in many lakes in central Florida are falling and require augmentation from wells to maintain surface levels. Wetlands adjacent to some irrigated lands are being degraded, if not completely eliminated. Water quality problems from agriculture are caused not by return flows as is the case in the West, but by stormwater runoff. Runoff from sugarcane and vegetable production in the Everglades Agricultural Area (EAA) is a leading concern of government agencies charged with the protection of the Everglades ecosystem. Even in areas where irrigation water supplies have not been limiting, concern over contaminants in runoff, especially nutrients, is leading to a reduction in farm acreage.

The large-scale environmental systems, which include not only the Everglades, but also the many estuarine areas that evolved under water-rich conditions, have become a dominant force in the debate over future water use. The goal of ecosystem restoration has become a direct limit on new water use in adjacent areas and is also being debated by government, industry, and environmental groups considering reallocation of water from existing uses to the environment. One of the most critical and controversial environmental issues in Florida centers on the nutrient enrichment of portions of the Everglades by stormwater runoff from the sugarcane and vegetable fields south of Lake Okeechobee. In the 1960s, some 500,000 acres of sawgrass prairie were transformed into the Everglades Agricultural Area by the federally authorized and constructed Central and Southern Florida Project. Currently, there are approximately 425,000 acres in sugarcane, 32,500 acres in vegetables, 12,000 acres in rice, and 25,000 acres in sod production. Vegetable farmers grow multiple crops, so the actual vegetable acreage harvested is closer to 70,000 acres. Most farms within the EAA are large, encompassing thousands of acres. Irrigation and drainage are provided by an on-farm network of canals connected to the federal Central and Southern Florida Project.

The Everglades evolved 5,000 years ago as an oligotropic (very low in nutrients) system. Today, stormwater runoff from the EAA is pumped directly into the remaining undeveloped Everglades. The water, while very low in phosphorus compared to other agricultural or urban runoff, contains phosphorus concentrations about 15 times the background levels of the marsh (150 versus 10 parts per billion). The runoff coming from the EAA is considered one of the contributing factors in the expansion of dense cattail growth into native sawgrass prairie systems.

The controversy over water quality problems in the Everglades spawned 5 years of bitter litigation between cane and vegetable growers and state and federal agencies. In 1994 the state passed special legislation outlining an interim approach to the problem—the construction over the next decade of 40,000 acres of artificial marshes to act as nutrient filters for the runoff. The cost to farmers is expected to be between $200 and $320 million over the next 20 years, and 25,000 acres will be removed from production. Federal resource agencies are in the early stages of considering proposals to remove another 100,000 acres of the most productive land from production in the next 10 years. This approach to solving the water quality problems of the Everglades has come with another significant cost. The years of expensive litigation have reduced the potential for collaborative efforts between the government, agriculture, and environmental groups.

A case with far less controversy centers on the expansion of citrus production into southwest Florida. Florida was hit in the early 1980s by a series of freezes that severely damaged production in the historic citrus belt in the center of the state. Since that time, citrus production has been moving south to avoid frost damage. Citrus acreage south of Lake Okeechobee has doubled in the past 10 years to 148,000 acres. Permitting for new groves continues, and the total irrigated area could climb another 50,000 acres by the year 2000 (Mazotti et al., 1992), although weakened market conditions may delay this process.

Historically, the southwest Florida citrus area consisted of wetlands (61 percent) and uplands (39 percent) dominated by pine flatwoods. By 1973, some 36 percent of the total area had been converted to agricultural use, first to pasture and then to crops and citrus. Today, 60 percent of the freshwater marshes and 88 percent of the pinelands have been lost. Although citrus groves do not necessarily eliminate biological diversity (Mazotti et al., 1992), the linkage between uplands and wetlands is critical to maintaining biological integrity. The fragmented remnant flatwoods are critical habitat for more wildlife species than any other cover type and are vulnerable to further development.

In response to the continuing loss of temporary wetlands, and the loss or fragmentation of forest and range habitats, the South Florida WMD is developing new rules to require a thorough evaluation of every new and existing water use to eliminate any detrimental effects on wetlands. Federal agencies are also requiring endangered species reviews on all major changes in upland areas. The citrus industry has responded quickly to these changes. Citrus farmers have been pioneers in the development of new technology for water conservation, and they have worked with regulatory agencies to find ways to preserve many habitat values. While they are certainly not immune from the environmental and competitive forces facing agriculture, they have not been confronted with the intense pressures facing farmers in the Everglades.

Water Supply Issues

In Florida, water conservation has become a necessity. In some areas, available supplies are limited by subsurface salt water intrusion; in other areas, supplies are limited by competing needs of nearby wetlands. There has been an aggressive initiative by agribusiness to develop the most efficient irrigation systems possible. There has also been an equally determined program by government institutions to encourage and, in some areas, mandate such technology shifts. In the mid-1980s, there was considerable focus on increasing irrigation efficiency.

Of the more than 4.6 million acres of commercial agricultural land in Florida, over 2 million acres (44 percent) are irrigated (Smajstrla et al., 1993). Farmers have begun to adopt more efficient irrigation technologies, including microirrigation. Currently, 418,000 acres are irrigated with microirrigation systems, and almost 94 percent of these acres are in fruit crops, primarily citrus. Approximately 50 percent of the current 2 million acres are adaptable and may be expected to convert to microirrigation. The rate of conversion is estimated to be about 31,000 acres per year, with most of this occurring in fruit and vegetable crops (Smajstrla et al., 1993).

In Orlando, 23 million gallons per day of reclaimed water is now being distributed to citrus groves for irrigation. The water, which has to meet rigorous water quality standards, is being used on 21 grove sites through 29 miles of pressurized distribution lines. To help meet the demands for citrus and turfgrass irrigation, and address the increased competition for water use, reclaimed wastewater for irrigation has increased from zero in 1970 to 51 million gallons per day in 1985 and to 170 in 1990.

The significance of water supply issues, specifically the competition between urban and agricultural water uses, can be seen in the example of the Tampa Bay region. In 1989, agricultural water use accounted for 64 percent of the total ground water withdrawn in the Floridan aquifer, the area's primary source of water, west and south of Tampa Bay. Citrus, tomatoes, and other vegetables make up the largest irrigated acreage in the area. Except for relatively short-term fluctuations caused by freezes, total citrus acreage has remained fairly constant at about 260,000 acres since the 1960s.

Continued use of the aquifer would result in salt water intrusion, permanent decline in lake levels, and the loss of wetlands. The water level in one of the most severely affected lakes has dropped 14 feet in the past 10 years. Over 90 lakes in the area require well water augmentation to maintain water levels (Bajwa, 1985). Test wells in Hillsborough and Sarasota counties have doubled in chloride content to 1,900 and 1,400 milligrams per liter, respectively. In response to these problems, the Southwest Florida WMD has stopped issuing new permits for ground water withdrawals until regulations requiring increased water use efficiency for all users can be implemented. Water-conserving technologies will be required for both new and existing users. Agricultural water use permits will be based on system efficiency, crop efficiency, and irrigation management (Southwest Florida Water Management District, 1993). The citrus industry, which has already installed microirrigation technology, is not expected to be affected. Tomato, melon, and potato farmers are expected to convert to fully enclosed seepage techniques or add drip irrigation.

A preliminary economic analysis commissioned by the Southwest Florida WMD found that the plan is not expected to significantly reduce the agricultural economy in terms of sales and employment through the year 2015. However, irrigators will be required to finance new water conservation technologies, which will lower business earnings. If growers maintain existing irrigation system efficiencies, total acreage in production will decline. Sod production is expected to shift to sprinkler systems to increase irrigation efficiency.

Conclusion

Although national statistics on the importance of irrigation are dominated by western states, Florida is ranked tenth in total irrigated acreage (2.1 million acres) and fourth in market value of irrigated crops harvested ($3.3 billion). Irrigated agriculture in Florida has grown substantially in the past decade and is projected to grow significantly over the next 30 years. Irrigation in the region relies heavily on ground water even though surface waters are extensive.

Competition for water is becoming intense, as is the pressure on irrigated agriculture from environmental regulation of water and land use. Tight restrictions on impacts to wetlands, and the desire to restore many previously disturbed natural systems, could severely limit future growth of irrigated agriculture, and in some cases may significantly reduce the number of acres in production. Agriculture has responded to these pressures with a more scientific approach to water use and wholesale conversion to new technology and management techniques. In some cases, though, the debate has included litigation, media warfare, and political skirmishing by both government and agriculture. In a few instances, pressure on agriculture has led to business failures and community hardship. The institutions that manage water have also changed, in some cases to try to solve these water problems through research and cost-sharing programs, and in others to use their regulatory power to force change on the irrigation industry.

With changes in the demographic composition of the state, and related changes in political leadership, traditional alliances and public support for agriculture are weakening. It will take years to rebuild the trust between agriculture and the government in the Everglades region. On the other hand, the long history of the citrus industry and the fact that it is not centered in or near the Everglades have nurtured a cooperative relationship between that industry and the government, one that is likely to endure. Ultimately, the future of irrigated agriculture in Florida will not be limited by the supply of water. It will depend on the ability of agriculture, urban water uses, and environmental interests to commit to a collaborative process of achieving mutually acceptable solutions to the state's water resource challenges. Recent experience indicates that when problems are addressed at the local level, with all stakeholders participating, lasting solutions are possible.

REFERENCES

American Fisheries Society. 1991. Pacific Salmon at the Crossroads: Stocks at Risk from California, Oregon, Idaho, and Washington, Vol. 16, No. 2, March-April. J. E. Williams and J. A. Lichatowich, eds.

Bajwa, R. S. 1985. Analysis of Irrigation Potential in the Southeast: Florida, A Special Report. Natural Resource Economics Division, Economic Research Service ERS Staff Report No. AGE851021. Washington, D.C.: U.S. Department of Agriculture, P. 42.

Banks, H. O., J. O. Williams, and J. B. Harris. 1984. Developing new water supplies. In Water Scarcity: Impacts on Western Agriculture. E. A. Englebert and A. F. Scheuring, eds. Berkeley: University of California Press. Pp. 109-126.

Beattie, B. R. 1981. Irrigated agriculture and the problems and policy alternatives. Western Journal of Agricultural Economics 7 (December 1981):289-299.

Bittinger, M. W., and E. B. Green. 1980. You Never Miss the Water Till. . . (The Ogallala Story). Resource Consultants, Inc. Littleton, Colo.: Water Resources Publication.

Bonneville Power Administration. 1993. Modified Stream Flows, 1990 Level of Irrigation, Columbia River and Coastal Basins, 1928-1989. Portland, Oregon: Bonneville Power Administration.

Bonneville Power Administration, U.S. Corps of Engineers, and U.S. Bureau of Reclamation. 1994. Columbia River System Operation Review, Draft Environmental Impact Statement and various appendices. Portland, Oregon: Columbia River System Operation Review Task Force.

Bryant, K. J., and R. D. Lacewell. 1988. Adoption of Sprinkler Irrigation on the Texas High Plains: 1958 to 1984. Department of Agricultural Economics, Texas Agricultural Experiment Station, Department Information Report DIR 88-1. College Station, Tex.: Texas A&M University.

California Department of Water Resources. 1994. California Water Plan Update. Bulletin 160-93. Sacramento, Calif.: Department of Water Resources.

Camp, C. R., E. J. Sadler, R. E. Sneed, J. E. Hook, and S. Ligetvari. 1990. Irrigation for humid areas. In Management of Farm Irrigation Systems. G. J. Hoffman, T. A. Howell, and K. H. Solomon, eds. St. Joseph, Mich.: American Society of Agricultural Engineers. Pp. 551-578.

Checcio, E., and B. Colby. 1993. Indian Water Rights: Negotiating the Future. Water Resources Research Center, University of Arizona.

Council for Agricultural Science and Technology (CAST). 1988. Effective Use of Water in Irrigated Agriculture. Report No. 113.

Demment, M. W., K. G. Cassman, W. J. Chancellor, E. W. Learn, R. S. Loomis, D. N. Manns, D. R. Nielsen, J. N. Seiber, and F. G. Zalom. 1993. California Farming System: Diversity to Compete in a Changing World. Agriculture Issues Center. Davis, Calif.: University of California.

Ellis, J. R., R. D. Lacewell, and D. R. Reneau. 1985. Economic Implications of Water-Related Technologies for Agriculture: Texas High Plains. Texas Agricultural Experiment Station MP-1577. College Station, Tex.: Texas A&M University.

Frederick, K. D., and J. C. Hanson. 1982. Water for Western Agriculture. Washington, D.C.: Resources for the Future.

Gilley, J. R., and E. Fereres-Castiel. 1983. Efficient use of water on the farm. OTA commissioned paper, excerpted in Water Related Technologies for Sustainable Agriculture in U.S. Arid/Semi-arid Lands. OTA-F-212. Washington, D.C.: U.S. Congress, Office of Technology Assessment.

Great Plains Agricultural Council, Water Quality Task Force. 1992. Agriculture and Water Quality in the Great Plains: Status and Recommendations. Publication No. 140. College Station, Tex: Texas Agricultural Experiment Station.

High Plains Associates. 1982. Six-State High Plains–Ogallala Aquifer Regional Resources Study. Report to the U.S. Department of Commerce and the High Plains Study Council. Austin, Tex.: High Plains Associates.

Idaho Agricultural Statistics Service. 1992. Idaho Agricultural Statistics. Boise: Idaho Statistical Reporting Service.

Kromm, D. E., and S. E. White. 1987. Interstate groundwater management preference differences: The Ogallala region. Journal of Geography 86(1):5-11.

Kromm, D. E., and S. E. White. 1992a. Groundwater problems. In Groundwater Exploitation in the High Plains, D. E. Kromm and S. E. White, eds. Lawrence, Kans.: University Press of Kansas. Pp. 1-27.

Kromm, D. E., and S. E. White. 1992b. The High Plains Ogallala region. In Groundwater Exploitation in the High Plains, D. E. Kromm and S. E. White, eds. Lawrence, Kans.: University Press of Kansas. Pp. 44-63.

Lacewell, R. D., and J. G. Lee. 1988. Land and water management issues: Texas High Plains. In Water and Arid Lands of the Western United States, A World Resources Institute Book. M. T. El-Ashry and D. C. Gibbons, eds., Cambridge and New York: Cambridge University Press. Pp. 124-167.

Lacewell, R. D., and E. Segarra. 1993. Farmers', Lenders' and Water Districts' Response to Texas Low Interest Loan Program for Water Conservation in Agriculture. Texas Water Resources Institute. TR-164. College Station, Tex.: Texas Agricultural Experiment Station, Texas A&M University.

Lacewell, R. D., J. R. Ellis, and R. C. Griffin. 1985. Economic efficiency implications of changing groundwater use patterns. In Issues in Groundwater Management, E. T. Smerdon and W. R. Jordan, eds. Austin, Tex.: University of Texas, Center for Research in Water Resources.

Lee, J. G. 1987. Risk implications of the transition to dryland agricultural production on the Texas High Plains. Doctoral thesis, Texas A&M University, College Station.

Mapp, H. P. 1988. Irrigated agriculture on the High Plains: An uncertain future. Western Journal of Agricultural Economics 13:339-347.

Marella, R. L. 1992. Water Withdrawals, Use and Trends in Florida, 1990. U.S. Geological Survey, Water-Resources Investigations Report 92-4140. Tallahassee, Fla. P. 38.

Mazotti, F. E., L. A. Brandt, L. G. Pearlstrine, W. M. Kitchens, T. A. Obreza, F. C. Depkin, N. E. Morris, and C. E. Arnold. 1992. An Evaluation of the Regional Effects of New Citrus Development of the Ecological Integrity of Wildlife Resources in Southwest Florida. West Palm Beach, Fla.: Southwest Florida Water Management District.

Moore, M. R., W. M. Crosswhite, and J. E. Hostetler. 1987. Agricultural Water Use in the U.S. 1950-1985. In National Water Summary 1987 - Hydrological Events and Water Supply and Use: USGS Water-Supply Paper 2350. J. E. Carr, E. B. Chase, R. W. Poulson, and D. W. Moody, compilers. Reston, Va.: U.S. Geological Survey.

National Research Council. 1989. Irrigation-Induced Water Quality Problems. Washington, D.C.: National Academy Press.

Northwest Power Planning Council. 1994. Columbia River Basin Fish and Wildlife Program. Portland, Oregon: Northwest Power Planning Council.

Opie, J. 1993. Ogallala: Water for a Dry Land. Lincoln, Neb.: University of Nebraska Press.

Reisner, M. 1993. Cadillac Desert. New York: Penguin Books.

Rosenberg, H. R., R. E. Garrett, R. E. Voss, and D. L. Mitchell. 1993. Labor and competitive agriculture technology in 1990-2010. In On the Brink of a New Millenium. Oakland: University of California, Agriculture Issues Center.

Smajstrla, A. G., W. G. Boggess, B. J. Boman, G. A. Clark, D. Z. Haman, T. A. Obreza, L. R. Parsons, F. M. Rhoads, T. Yeager, and F. S. Zazueta. 1993. Microirrigation in Florida: Systems, Acreage and Costs. Bulletin 276. Florida Cooperative Extension Service, University of Florida. P. 12.

Southwest Florida Water Management District. 1993. Southern Water Use Caution Area Management Plan. Draft 9-01-93. West Palm Beach, Fla: Southwest Florida Water Management District. P. 98.

Steinbergs, C. Z. 1994. Retrofitting a golf course for recycled water. An engineer's perspective. In Wastewater Reuse for Golf Course Irrigation. Chelsea, Mich.: Lewis Publishers.

Stewart, B. A., and W. L. Harman. 1984. Environmental impacts. In Water Scarcity: Impacts on Western Agricuture, E. A. Engelbert and A. F. Scheuring, eds. Berkeley, Calif.: University of California Press.

U.S. Department of Agriculture. 1989. The Second RCA Appraisal: Soil, Water, and Related Resources on Nonfederal Land in the U.S. Washington, D.C.: U.S. Department of Agriculture.

Williford, G. H., B. Beattie, and R. D. Lacewell. 1976. Effects of a Declining Groundwater Supply in the Northern Plains of Oklahoma and Texas on Community Service Expenditures. Texas Water Resources Institute TR-71. College Station, Tex.: Texas A&M University.

Zwingle, E. 1993. Ogallala Aquifer: Wellspring of the High Plains. National Geographic. March: 80-109.

6

Future Directions

Irrigation, to use a hydrological metaphor, is at a watershed divide. The use of specially provided water enables the production of food, fiber, and landscaping at levels and in places that would not otherwise be possible. But in recent years the public has grown increasingly concerned about the economic and environmental costs of irrigation. Some people see the dedication of a substantial portion of available water supplies to irrigation as inequitable and inefficient. This is especially true in the more arid regions of the country where water supplies are limited and competition from other water users is increasing. In the face of ever larger federal budget deficits, the financial costs of providing support to irrigation also loom large. In addition, the environmental consequences of irrigation—measured in terms of water diversions and consumption, impacts on water quality, and effects on aquatic, plant, and wildlife habitats—are increasingly being questioned.

Irrigation is a means to an end. It is a valuable tool, rooted in ancient tradition, that has proven to be dynamic and flexible. To a considerable extent, the future of irrigation depends on our ability to find ways to use this tool in a manner that continues to provide important benefits but with fewer and more acceptable environmental and economic costs.

Over the course of this study, the committee has examined many factors that may affect the future of irrigation. These factors—competition for water; concerns over environmental impacts, including the potential impacts of climate change; continued urbanization; conservative fiscal policy; the globalization of the United States economy; the shifting roles of federal and state governments; and tribal economic development—will affect irrigation differently in different regions. The following discussion summarizes this report's primary conclusions and suggests some likely future directions for irrigation in the United States.

Conclusion 1

Irrigation will continue to play an important role in the United States over the next 25 years, although certainly there will be changes in its character, methods, and scope. It is likely that total irrigated acreage will decline, but the value of irrigated production will remain about the same because of shifts to higher-value crops.

Total irrigated land in farms will decline from the 1994 total of 52 million acres, but it is expected that irrigation will continue to account for roughly the same percentage of the total value of agricultural production. Some important regional and farm-level variability underlie these predictions, however. For instance, regions dependent on declining ground water supplies are likely to experience continued declines in irrigated acreage. A few regions will see continued growth of irrigation, such as the lower Mississippi Valley and the Southeast. Overall, the total value of irrigated production may change little as yields increase and land is planted with higher-value crops. Successful farmers will adapt to increased water scarcity, new requirements to protect water quality and maintain streamflows, reduced crop and water subsidies, and global agricultural markets through innovation in technology, management, and marketing strategies. They also may have to adapt to higher energy costs and regional climatic changes associated with any global warming.

Future Directions:

- *Competition among water users is increasing. The impacts of large-scale irrigation in the arid West on the environment and the availability of water for other uses will command continuing, and perhaps increased, policy and management attention by the federal, state, and tribal governments. In the West, attention will need to be focused on: physical and economic efficiency of irrigation systems and water use; implementation of American Indian water rights; the expansion and administration of water markets and transfers; environmental impacts; and institutional reform. By contrast, irrigation policies and institutions in humid regions are less well developed and will need to evolve to reflect contemporary and future irrigation issues. In short, policymakers will have to focus primarily on the quality of irrigation in the West and on the expansion of irrigation in the East.*

Conclusion 2

Given changing societal values and increasing competition for water,

the amount of water dedicated to agricultural irrigation will decline. The availability and cost of water to the farmer are likely to remain the principal determinants of the extent of irrigation in the western United States; they are becoming increasingly important influences in the southern and eastern states as well.

Irrigation, particularly for agriculture, accounts for more withdrawals of water from surface and ground water sources and more consumption of water than any other use in the United States. With increasing pressures on limited water supplies, existing water uses necessarily are subjected to greater scrutiny. The use of water for agricultural irrigation and landscaping will continue to be vitally important, but as demand for water for other uses increases the need to allocate water to these uses will cause the amount of water dedicated to irrigation to decline. In particular, an increasing share of water supplies will be used for urban and environmental uses.

This shift in water use already is occurring through multiple processes, including market-based transfers of water, more efficient use of water in irrigation, and changes in the operation of water storage and delivery facilities. Voluntary transfers are being used more frequently to meet new urban demands and, in fewer cases, to meet environmental demands. Increased efficiency is most common where the irrigator's cost of water is increasing, or where historically available supplies are reduced or irrigators are able to sell or otherwise profit from conserved water. Historically, conservation and technological innovation in the West lagged because of subsidies and institutional arrangements that kept the cost of water to farmers artificially low. From a technical and management perspective, there are many ways in which irrigation can be made more efficient. There are few incentives at present, however, for water users to make the necessary investments in efficiency improvements. Similarly, there are many ways in which water storage and delivery systems can be operated to provide benefits to a larger number of uses and users. The incentives to make these changes have only recently begun to emerge. Overall, current institutions (e.g., state and federal water allocation and pricing policies) are not ideally structured to ease the transition that must occur.

In regard to water transfers, the committee found some confusion and misconceptions regarding the distinction between water withdrawals and water consumption. This can be a serious impediment to states and water agencies or districts striving to establish effective water conservation programs and reallocation strategies.

Future Directions:

- *New irrigation uses will be expected to meet increasingly strict standards of efficiency as a condition of use. Over time, existing uses will experience increasing pressure—in the form of prices, regulation, or incentives—to increase irrigation efficiency as well. Historical practices that place too great a burden on available water supplies or water quality will be legally challenged as being "unreasonable" or wasteful. In response, states will need to clarify and revise laws governing the rights to control and/or transfer conserved water, including elements designed to foster mitigation of third-party effects.*

- *States will need to establish improved systems to facilitate the voluntary transfer of water. These systems should provide clear rules and well-defined processes by which transfers can occur and should incorporate measures for protecting other existing uses of water. Such measures would include criteria for quantifying the amount of water that can be transferred according to predefined standards regarding consumptive use and system delivery efficiencies. These criteria should not unduly limit incentives to transfer water rights. The transfer process may be managed at the local or regional level to more effectively address water user and third-party concerns, under rules established at the state level.*
- *Each state should carefully review its definition of water availability. It is essential that consumptive use of water be the unit of measure, as is the case, for instance, in California. Similarly, it is essential that accurate and consistent definitions of water conservation be developed to further the implementation of new technologies. Often, a new irrigation technology or practice that improves water use efficiency does not reduce consumptive use proportionally because drainage or leaching factors are not known or the conserved water is shifted to irrigate more acres. Overallocation caused by not accurately estimating and considering consumptive use of water will only increase the potential for water conflicts in the future.*

Conclusion 3

The economic forces driving irrigated agriculture increasingly will be determined by our ability to compete in global markets. This shift toward globalization, combined with reductions in protection and support for individual farmers, means that farmers will have to deal with increased levels of risk and uncertainty.

With the negotiation of the NAFTA and GATT trade agreements, goods produced by irrigated agriculture increasingly will compete in an international market environment. Crop support prices will decrease (and perhaps end, if current proposals are enacted), and the costs of water supplies and environmental compliance will increase, forcing irrigated agriculture as a whole to move toward a more competitive structure. Producers of some crops, particularly higher-value crops such as vegetables, will face greater competition from foreign producers. Individual farmers will be more vulnerable to the vagaries of weather. Because of the higher capitalization necessary for irrigated farms, irrigators also are more vulnerable to fluctuations in crop and energy prices. In response, some of the production of high-value crops will shift to regions with more reliable water supplies or other advantages. In general, the trend toward globalization will provide better market opportunities for some crops and increased competition for others. The potential economic rewards to irrigated agriculture may be significant, but economic risk from increased competition will also increase.

Future Directions:

• *To tap international markets, farmers will need different skills, communication modes, and information. To help farmers compete effectively in these markets, educational systems—including research and teaching, will need to evolve. These systems will need to be innovative in integrating, interpreting, and disseminating information; be more global in perspective; and recognize the intense pressures associated with operating irrigated farms in the modern economic and environmentally conscious context.*

• *In a highly competitive world market, it is critical that U.S. farmers are not put at a comparative disadvantage. Products imported to the United States should meet the same food safety and chemical use standards required of U.S. producers. The Department of Agriculture must be effective in enforcing international agreements.*

• *To improve marketing efficiency and reduce risk, federal and state agencies need to develop education programs and work with commodity groups on market promotion and risk reduction tools. Also, accurate market data are essential for U.S. producers to compete effectively, and the Department of Agriculture will have a critical role in crop and market data development.*

Conclusion 4

The structure of irrigated agriculture will continue to shift in favor of large, well-financed, integrated, and diversified farm operations. Smaller, under-financed operations or those with less-skilled managers will tend to decline.

The long-term trend toward vertical and horizontal integration in farm operations will continue. The number of farmers declined from 6.45 million in 1920 to 1.93 million in 1992, a decline of 70 percent. With changes in U.S. farm policy leading to less federal control and subsidies along with a much greater emphasis of global market influences, the consolidation toward fewer and larger farms is expected to accelerate. Market risks are going to increase and irrigators will have incentives to increase in size to take advantage of economies of scale as well as to reduce risks by integrating into processing and marketing. Larger integrated farm operations will be better positioned to benefit from new technology and information.

Concurrent with this trend toward larger well-financed and integrated farm operations, there will be growth in small specialty farms located near urban markets. These farms will tend to be labor intensive but produce high-value crops for a niche market (e.g., organic foods, gourmet restaurants, fresh fruits, and vegetables delivered to the urban consumer). The total volume of production from these farms will be a very small percentage of total production from irrigated agriculture.

This suggests that the most vulnerable group of irrigators is the medium-sized farmers, especially those that are heavily leveraged. Thus the overall nature of irrigated farms will move toward a bimodal distribution: very large integrated irrigated farms producing the vast majority of output and very small specialty farms. Farmers who relied on depreciation to survive may lack the capital and borrowing capacity to overcome serious commodity price swings.

Future Directions:

- *Medium-sized farms will face particular challenges in the increasingly global, competitive environment. Some strategies that might enhance survival of these operations would involve cooperation among local or regional groups of farmers to (1) pool products to achieve greater market impacts, (2) arrange large-quantity purchasing of inputs for quantity discounts and (3) develop methods to share the use of very expensive equipment and market information. This approach brings some loss of independence, but provides tools for survival in a more competitive marketplace. The precedent for this organization structure is seen in processing and marketing producer cooperatives. Technical assistance to develop and organize a formal farmer cooperative, as well as a strong financial commitment from the farmers, is essential.*

Conclusion 5

Many important federal, state, and local policies and institutions affecting irrigation were established in a different era and they no longer meet contemporary societal needs. Changes in these policies and institutions are occurring to reflect changing economies, emerging values, and shifting policy priorities. Thus, for example, the Bureau of Reclamation is moving from a project construction agency to a water management agency. Innovation and flexibility will be needed, especially as direct federal support continues to diminish.

As discussed in Chapter 4, much of the policy and institutional development in support of irrigation occurred long ago. The basic principles of western water law governing the allocation and use of water from surface water resources were well established by the turn of the century. Federal reclamation policy took root shortly thereafter. Taken together, the purpose of much of federal and state water policy is intended to encourage irrigated agriculture in the arid western states.

The support of irrigation remains an important federal and state policy objective, but it is no longer the driving public policy objective in the West. Indeed, it is not uncommon today for irrigation, as it has been historically practiced and supported, to conflict with other important policy objectives such as reducing the federal budget deficit, encouraging recreational uses of water, and protecting water quality. The result is a complex array of incentives and penalties that confuse irrigators and sometimes put government agencies and policies at odds with one another.

In part, this confusion reflects a period of transition. The consumers of low-cost food and green lawns want the benefits of irrigation, but object to some of the costs—both financial and environmental—that must be paid. There is a clear need to revisit earlier policy and institutional choices in a search for ways that can continue to provide the desired benefits of irrigation but at more acceptable costs.

Future Directions:

- *Federal policies supporting the provision of below-cost water will gradually change to better reflect the full costs of making water available for different users.*
- *The federal role in the supply of water for irrigation and other uses will decline through time as local and regional entities take over more of these responsibilities.*
- *Uncertainties over water rights impede effective allocation of water resources. In the West, resolution of tribal water rights claims and rights to market water for off-reservation use would remove a major source of uncertainty. The federal government, in the continuing exercise of its trust responsibility to Indian tribes, will be challenged to commit significant attention to settling tribal claims to water and helping the tribes realize the benefits of their water rights.*
- *Federal requirements related to such things as protection of endangered species and water quality will continue, but the means by which these objectives will be pursued will shift to allow more flexibility in the regulatory programs, to encourage the use of incentives and market-based approaches, and to engage more local, regional, and state participation in their achievement.*
- *States will continue to revise and amend water allocation law in a manner that emphasizes more efficient use of water to accomplish beneficial purposes, supports a broad range of water uses including instream and other environmental uses, and broadens the authorities of irrigation districts to include other water-based interests within the district areas.*

Conclusion 6

In the past, the term "irrigation" effectively meant irrigation for agriculture. But the nature of irrigation has changed dramatically in the past

two decades and will continue to change. Turf irrigation is now an important part of the irrigation industry, and irrigation for urban landscaping and golf courses in particular will continue to expand as urban populations increase.

Urban land area will continue to grow, and with them the demand for landscaping, trees, lawns, and golf courses. Urban users typically can better afford the cost of water and other costs of irrigation, so they will continue to compete with agriculture for access to water supplies. Some policy changes may prove necessary to encourage efficient water use, such as tiered water pricing programs that charge higher prices for incremental increases in water use. As water costs rise, interest in the use of native or drought-tolerant plants and other landscaping techniques that require little supplemental water will be encouraged.

Future Directions:

- *Urban landscaping provides an important opportunity to develop and expand the use of water reuse systems or other forms of wastewater reclamation. Experience to date in applying reclaimed water in the irrigation of parks, median strips, and golf courses suggests that treated wastewater can be used safely and effectively in these ways, at lower cost and with less demand on freshwater supplies.*
- *Intensive urban landscaping is becoming a significant source of ground water contamination, primarily as a result of the overapplication of pesticides and fertilizers. Research and increased awareness of these problems can bring about more effective requirements for controlling this pollution. Cities and urban water districts will need to play a more active role in education and demonstration about appropriate landscape practices.*
- *Developments in water management for landscaping will have some applications in agricultural irrigation, just as applications developed for agriculture have been adapted for landscape use. The landscape and agricultural irrigation sectors should work together to ensure that improvements in irrigation efficiency and water reuse have the broadest possible impact. Improved communication can be facilitated by the private sector as it develops and markets methods for advanced irrigation scheduling, nonpoint-source pollution control, and irrigation technologies.*
- *Water prices remain low in many locations, but increased competition for urban water supplies will lead to higher prices in the future for all users. Urban water supply organizations increasingly should adopt rate structures intended to encourage more efficient use of water.*

Conclusion 7

Advances in irrigation technology are necessary if both agricultural and

turf irrigation are going to adapt to changing demands and changing supplies. The irrigation industry will need to play a larger role in technology development and dissemination as the federal government trims its support for these activities.

In both agriculture and turf irrigation, there has been a significant shift toward the use of sprinkler irrigation and microirrigation technologies. Surface irrigation is being modernized in most regions with the use of laser leveling of land and automated systems such as surge irrigation. Electronic controls and sensors improve the control and management of irrigation systems. The forces promoting the adoption of new technologies are the increasing costs of labor, energy, and water as well as limitations on water availability. Constraints to the adoption of more efficient technologies are the costs to purchase and install systems, the lack of management skills, and inadequate incentives. Often, a threshold level of capital and size is necessary to take advantage of technological advances. Advances in plant genetics may offer some possibility for reducing water requirements for some crops. To date, however, progress in this area has been limited.

Future Directions:

- *The pressures for greater efficiency will increase and will require the development and transfer of new technology. The federal government has supported much of the research and development of new irrigation technologies. With the continuing need to reduce federal expenditures, more leadership and funding for this research will have to come from the private sector and through partnerships between irrigators, the private sector, and state and federal researchers. Manufacturers should increase their research efforts, and agriculture and turf irrigators should increase support for this research. In addition, irrigation districts and similar organizations should become more active in encouraging the testing and demonstration of new technologies and in educating irrigators to use cost-effective technologies.*

Conclusion 8

Some portion of the water now in agricultural use will over time be shifted to satisfy environmental goals. In addition, there will be continued pressure to reduce environmental problems associated with irrigation—both agricultural and turf.

Irrigation has numerous environmental effects, some negative and some positive. For example, irrigation produces vegetation where it would not normally exist. In turn, that vegetation can provide aesthetic benefits, recreational opportunities, and valuable wildlife habitat. Return flows from irrigated lands can help to maintain flows in some streams during periods in which natural flows would be much lower. Storing streamflows in reservoirs during times of surplus allows the release of this water at times when it would not otherwise be available for environmental and recreational uses.

On the other hand, irrigation degrades water quality and has significant water supply impacts. The storage and diversion of water for irrigation profoundly alter the natural hydrology of streams and the habitats of native plants and animals that depend on them. The application of large quantities of water to irrigated lands results in soil erosion and the sedimentation of streambeds and spawning gravels. In addition to sediments, salts and other constituents are leached from the soil and transported into rivers and streams. Some of these constituents, such as selenium, pose significant threats to fish and wildlife. Fertilizers and pesticides used in irrigated agriculture contaminate both surface water and ground water systems.

In the past, federal and state laws have tended to exempt irrigation from legal responsibility for some of its adverse environmental effects. A prominent example is the explicit exemption from regulation of irrigation return flows under the Clean Water Act. There is growing public recognition, however, that agriculture, including irrigated agriculture, is a major source of water quality problems nationwide. To date, federal and state agencies have relied on demonstration projects and voluntary best management practices to address pollution caused by irrigation. Although nonregulatory approaches have brought some progress, stronger measures are likely to be needed to reduce water quality problems.

Future Directions:

- *The trend in environmental policy to regulate activities that affect endangered species, wetlands, water quality, and public health will continue, although it is likely to proceed at a slower pace than during the 1970s and 1980s. Given the diverse and variable nature of the environmental impacts of irrigation, federal, state, tribal, and regional agencies should look for regulatory and management approaches that can be tailored to specific problems and locations. Incentive-based programs, investment credits, point- and nonpoint-source trading programs, and other mechanisms provide this flexibility. The need for local- and regional-level environmental problem solving is consistent with the emerging support for watershed- and ecosystem-based initiatives.*
- *The availability and quality of water are limiting factors for irrigation and for ecosystems. Thus the future will inevitably bring opportunities for either conflict or cooperation between irrigation and environmental interests. Locally based programs for addressing environmental impacts, even those that employ incentives, require a high level of cooperation from stakeholders. Government agencies must participate as partners as well as regulators. In addition, processes to resolve these issues will require a high level of commitment and participation on the part of irrigators and the organizations that support them.*

Conclusion 9

Irrigation emerged as an individual and collective effort at the watershed level, and in many important respects its future will be determined in the local watershed.

Irrigation is greatly influenced by forces operating at the national and even global level but, at core, irrigation is a local activity. It was irrigation that made possible settlement of large portions of the arid western states, initially along creeks and rivers with adjacent bottomlands that could be readily provided with water. These small-scale, local efforts eventually evolved into large-scale, sometimes inter-basin water development projects served by complex hydraulic infrastructures maintained and run by large bureaucracies. As a result, irrigation became more an end in itself, rather than a means to an end. In the process, irrigators became increasingly influenced by state and federal policy. As water-related problem solving returns increasingly to the state and regional level, often organized around watersheds, irrigation interests need to reintegrate their goals into these more local processes.

We are generally enthusiastic about the reemergence of local initiatives in water matters. The now well-documented growth of watershed councils and other such entities reflects a potentially promising trend in water management. At the same time, watershed management poses challenges for existing institutions and agencies at the federal, state, and local levels whose jurisdictions and authorities are not necessarily designed to work within this kind of framework.

Future Directions:

- *Irrigators should become actively engaged in local watershed initiatives as a means by which accommodation of interests with those of others in the watershed may be reached. Often, watershed efforts are focused on changing the manner in which water is used. Irrigation is, in many instances, the dominant out-of-stream use of water and is, thus, likely to be a central focus of conflicts and negotiations over changing water demands and priorities. Understandably, irrigators may view watershed initiatives as threatening. Alternatively, watershed initiatives can be viewed as a forum in which irrigators and nonirrigators can gain a better appreciation and understanding of their respective goals and interests. We encourage irrigation interests to see watershed-based activities as an opportunity rather than a threat.*
- *Federal and state agencies should facilitate problem-solving at the watershed level by tailoring policies and programs to encourage such approaches, particularly where the issues can best be addressed at this level. In some cases this may mean providing increased flexibility in the manner in which explicit national or state objectives can be accomplished. It may require some reorganization of responsibilities among federal and state agencies. It is likely to require*

at least some financial and technical support for more locally driven watershed activities.

Final Thoughts

Irrigation made possible the settlement and development of a large portion of this country. Irrigation, and irrigation policy, provided the means by which people could transform land with little apparent economic value into productive farms and ranches. These irrigated lands continue to support the people who live on them and to produce products that support large numbers of other people. For many whose livelihoods depend on irrigation, it is far more than just a job—it is a way of life.

Irrigation is in a time of transition, which brings uncertainty and anxiety. But irrigators in the United States have demonstrated creativity and resilience in responding to significant changes over the decades (sometimes willingly and sometimes less so) and these characteristics will be critical in the coming years. The challenges confronting irrigation occur within the larger context of history. This context, or culture, of irrigation and the traditions out of which it arose attest to the innovative and adaptive nature of irrigation as a whole. It shapes the thinking of those who are part of the irrigation community, particularly those living in rural agricultural areas. It affects the manner in which people within the irrigation community understand the forces of change discussed in this study, and it clearly affects how they are responding to these changes.

The forces of change evident today and discussed in this study can be, and in many cases are, viewed by irrigators as threatening to this way of life. These forces challenge the value of irrigation by questioning whether and how much of this use of land and water is still justified. The committee rejects the view that irrigation is a Faustian bargain—a price too great to pay for the benefits it produces. Irrigation has served this nation well and will continue to do so in the future. But there will be changes in where and how irrigation occurs, particularly in the West. Some irrigators will prove unable to compete under the new conditions and will fail; others will see opportunity in change and thrive. Although the committee does not foresee explosive change on the horizon, certainly the future will bring surprises, some perhaps dramatic. It is critical that the same resourcefulness that has made irrigation such an important economic and cultural activity over the past 100 years be brought to bear in the future. If so, irrigation as a whole will likely continue to adopt more sustainable practices, practices that provide both economic and social benefits while reducing environmental harm.

Appendix A

Biographical Sketches of Committee Members

WILFORD GARDNER (Chair), is Dean Emeritus, College of Natural Resources, University of California at Berkeley. Dr. Gardner received M.S. and Ph.D. degrees in physics from Iowa State University in 1953. He has served as a physicist with the U.S. Salinity Laboratory, USDA; professor of soil and environmental physics, University of Wisconsin; and head of the Department of Soil and Water Science at the University of Arizona, Tucson. He has been a National Science Foundation Fellow at Cambridge and Wageningen and a Fullbright Fellow at Ghent University. Dr. Gardner is a member of the National Academy of Sciences and the Water Science and Technology Board. His research has centered on the state and movement of water and solutes in the vadose zone, soil-water-plant relations, and the kinetics of soil microorganisms.

KENNETH FREDERICK (Vice Chair) is a senior fellow at Resources for the Future (RFF). He received a Ph.D. in economics from the Massachusetts Institute of Technology in 1965 and a B.A. from Amherst College in 1961. Dr. Frederick has been a member of the research staff at RFF since 1971 and served as director of its Renewable Resources Division from 1977 to 1988. Prior to joining RFF, he served on the economics faculty at the California Institute of Technology and as an economic advisor in Brazil for the Agency for International Development. Dr. Frederick's recent research and writings have addressed the economic, environmental, and institutional aspects of water resource use and management. He is the author, coauthor, or editor of seven books and the author of more than 50 published papers dealing with these and other natural resource issues. He is a former member of the WSTB and served on the Board's Committee on Climate Uncertainty and Water Resources Management.

HEDIA ADELSMAN is a special assistant for land use planning and regulation with the Department of Ecology, State of Washington. She works on issues of environmental protection, land use planning, and growth and development. Previously, she was the manager of the Water Resources Program, Department of Ecology. She was responsible for leadership of all water resources management issues. She holds a B.A. in agricultural engineering from the University of Tunis, Tunisia, an M.A. in agricultural economics from the University of Minnesota, and an M.B.A. from the College of St. Thomas, Minnesota.

JOHN S. BOYER received an A.B. in biology in 1959 from Swarthmore College, Swarthmore, Pennsylvania; an M.S. in plant physiology from the University of Wisconsin in 1961; and a Ph.D. in plant physiology from Duke University in 1964. His research interests are metabolic mechanisms of losses in plant growth under dehydrating or saline conditions. His research explores photosynthesis, cell enlargement, and reproduction beginning at the level of the whole plant but using methods in biophysics, biochemistry, and molecular biology. The overall goal is to understand how growth is inhibited and whether it may be recovered. Experimental material includes agronomic species and marine plants in an effort to extend findings to practical applications. Since 1987, Dr. Boyer has been Dupont Professor of Marine Biochemistry/Biophysics, College of Marine Studies, University of Delaware. Dr. Boyer is a member of the National Academy of Sciences.

CHELSEA CONGDON is a water resource analyst with the Environmental Defense Fund (EDF) in its Rocky Mountain office. Ms. Congdon focuses primarily on issues of water resource management and policy in the Rocky Mountain region and in California. She works on issues related to the reform and improvement of water management practices in major river basins in the West; water allocation and water transfer policies at the state, tribal, and federal levels; agricultural nonpoint-source pollution control; and protection and restoration of aquatic ecosystems. Ms. Congdon is the coauthor of a study on the use of incentive-based approaches for addressing agricultural drainage problems in California's Central Valley. She has served as a board member of the California Irrigation Institute. She received her B.A. from Yale University in 1982 in resource policy and economics. In 1989, Chelsea completed a master's degree in energy and resources at the University of California at Berkeley with emphasis on state and tribal cooperation in water quality regulation.

DALE F. HEERMANN is an agricultural engineer and research leader for the Water Management Research Unit at Fort Collins, Colorado. He is an international authority on irrigation systems and irrigation scheduling technology. Dr. Heermann developed a technique for theoretically determining application depths, rates, and uniformities for center pivot sprinkler irrigation systems, which is of considerable importance to irrigation technology. He extended this work to the study of center pivot pressure distribution formulation and to the relationships of application rates to the intake rates of soil. He received his B.S. in agricultural engineering, 1959, University of Nebraska; M.S., agricultural engineering, 1964, Colorado State University; and Ph.D., 1968, Colorado State University.

EDWARD KANEMASU is Director of International Agriculture and Regents Professor at the University of Georgia at Athens, Georgia. Previously, he was head of the Department of Crop and Soil Sciences at Georgia and professor of agronomy and laboratory leader for the Evapotranspiration Laboratory, Kansas State University. He received his Ph.D. in environmental physics from the University of Wisconsin-Madison and his B.S. in soils and M.S. in soil physics from Montana State University. He is an expert on the water use efficiency of agronomic crops, evapotranspiration, and agricultural climatology. He was a member of CAST's task force on Water Use in Agriculture: Now and for the Future, the Science Advisory Panel for NASA on Global Habitability, and chair of the Great Plains Committee on Evapotranspiration.

RONALD D. LACEWELL received a B.S. in agricultural economics in 1963 from Texas Tech University, an M.S. in agricultural economics in 1967 from Texas Tech University, and a Ph.D. in agricultural economics in 1970 from Oklahoma State University. He is currently professor at Texas A&M University and Chairman of the Environmental Affairs Team of the Texas A&M University System Agricultural Program, which is responsible for coordinating environmental and natural resources research and education programs. He is also a delegate to the University Council on Water Research for the University. Current water-related research includes the impact of agricultural production systems on water quality, optimal waste management strategies and marketing alternatives for confined animal feeding operations to protect water quality, integrated pest management systems, and optimal irrigation strategies for crop production in west Texas.

LAWRENCE J. MacDONNELL holds degrees from the University of Michigan, (B.A. 1966); University of Denver College of Law (J.D., 1972); and Colorado School of Mines (Ph.D., 1975). He is a lawyer and consultant with Sustainability Initiatives in Boulder, Colorado. Between 1983 and 1994 he was the director of the Natural Resources Law Center at the University of Colorado School of Law. During this time he taught courses in water law, public land law, oil and gas law, and mining law. He served as principal investigator for 19 funded research projects with grants from seven different foundations and six different government agencies. He authored more than 50 publications, including books, law review articles, journal articles, and research reports. He has taught at the Colorado School of Mines, the University of Denver, and with the Colorado Outward Bound School.

THOMAS K. MacVICAR is president of a private consulting firm specializing in the water resource and environmental issues of south Florida. Prior to beginning his consulting practice in 1994, Mr. MacVicar spent 16 years on the staff of the South Florida Water Management District. From 1989 to 1994 he was the second in command of the 1,500 employee agency with direct responsibility for all water resource issues. He was the agency's chief negotiator and spokesperson for Everglades issues and had direct supervisory responsibility for the Planning, Regulation, Research, and Operations Departments. He is a member of the Florida Engineering Society and the American Society of Civil Engineers. He was the recipient of the 1987 Palladium Medal for Outstanding Engineering Achievement in the Support of Environmental Conservation, given jointly by the National Audubon Society and the American Association of Engineering Societies; and he received the national Marksman Award for Engineering Excellence given by the Engineering News Record Magazine. He earned his master's degree in water resource engineering from Cornell University and completed his B.S. in agricultural engineering at the University of Florida. He also received a B.A. in political science from the University of South Carolina.

STUART T. PYLE is a consulting civil engineer with experience in water resources. He retired from the Kern County Water Agency in 1992 after 17 years as general manager and 2 years as senior advisor. His professional career in water resources began with the California Division of Water Resources in 1948. He participated in the planning and development of numerous water management projects, including drafting the California Water Plan. At Kern County Water Agency, Mr. Pyle directed and managed the 1 million acre-foot share of the State Water Project. He is a consultant for the Kern County Water Agency. He represents the agency on a number of statewide organizations and is serving on the Department of Water Resources Advisory Committee for its 1993 update of the California Water Plan. Mr. Pyle has a bachelor's degree in civil engineering from Marquette University in Wisconsin.

LESTER SNOW (through February 16, 1995) received a B.S. in earth sciences in 1973 from Pennsylvania State University and a M.S. in water resource administration from University of Arizona in 1976. Mr. Snow has extensive experience related to the water needs of cities and the trade-offs inherent in a changing agricultural environment. He served as General Manager, Arizona Department of Water Resources, 1981-1987; Director, Tucson Active Management Area, 1984-1987; and Deputy Director, Tucson Active Management Area, 1984-1987; and Deputy Director, Tucson Active Management, 1981-1984. He was instrumental in making San Diego a leader in the water conservation movement and has received numerous awards for the excellent contributions made to the education process from a purveyor. He was with the San Diego County Water Authority from January 1988 to February 1995. Mr. Snow is currently executive director of CALFED Bay-Delta Program in Sacramento.

CATHERINE VANDEMOER is a water rights specialist with the Office of the Assistant Secretary, Indian Affairs, U.S. Department of the Interior. She received her Ph.D. in watershed management from the University of Arizona and a bachelor's degree in geology from Smith College. Previously, she was executive director of the Wind River Environmental Quality Commission, a resource manager and water engineer for the Wind River Reservation, and a hydrologist with the Council of Energy Resource Tribes. She was also owner of Watershed Management Systems, Oakland, California; the director of the Water Resources Program, American Indian Resources Institute; and a research coordinator for the Papago Water Survey at the John Muir Institute for Environmental Studies. Her research interests include the interface between conservation policies and actual conditions with respect to desertification and water quality and supply, integrated resource management, and federal stewardship responsibilities on Indian lands.

JAMES WATSON received his B.S. in agronomy at Texas A&M University in 1947 and his Ph.D. at Pennsylvania State University in 1950. Dr. Watson has conducted research on adaptability of species and cultivars of turfgrass; fertilization practices; irrigation and compaction effects on fairway turf; snowmold prevention; techniques for the winter protection of turfgrasses; and similar studies. He is contributor to several texts on turfgrass science, as well as author of well over 400 articles on turfgrass care and management, water conservation, cultural practices, and other areas of interest to the green industry. Dr. Watson was assistant professor in the Department of Agronomy at Texas A&M University. He joined the Toro Company in 1952 as Director of Agronomy. In 1985, Dr. Watson was elected director to the Boards of the Freshwater Foundation, Mound, Minnesota, and the National Golf Foundation, and in 1986 was selected as Landscape Management's Man of the Year and later that same year was chosen Man of the Year for *Landscape and Irrigation* magazine. In 1988 he was elected to the Board of the Sports Turf Managers Association. In 1994 Dr. Watson was presented the Donald Ross award by the American Society of Golf Course architects, and, in 1995 the Old Tom Morris Award by the Golf Course Superintendents Association of America. Also in 1994, Dr. Watson served as Agronomic Coordinator for World Cup Soccer Venues.

JAMES L. WESCOAT, Jr., earned his Ph.D. in 1983 and M.A. in 1979 in geography from the University of Chicago, and a B.L.A. in 1976 in landscape architecture from Louisiana State University. Dr. Wescoat currently is associate professor of geography at the University of Colorado at Boulder. His current research is on long-term water development in South Asia and the American West, with special emphasis on the geographical interactions between those regions that have shaped current water management problems. He has received fellowships and awards from the National Science Foundation, the Rockefeller Foundation for the Humanities, and Dumbarton Oaks.

HOWARD A. WUERTZ operates a farming venture that involves 2,360 acres of land in the Coolidge–Casa Grande area. This is a diversified operation devoted to cereal grains, cotton, seedless watermelons, and various other vegetable crops. He has been instrumental in the development of the River Cooperative Gin and Arizona Grain, Inc. He has worked for many years in the Farm Credit System and has been actively involved at the local, state, and federal levels on conservation and resource issues involving agriculture. Mr. Wuertz has pioneered the development of a subsurface drip irrigation system for use on cotton, grains, and other desert irrigated crops. This system has allowed for water savings of up to 50 percent while increasing yields and improving the quality of marginal soils. It has also necessitated the development of special machinery for minimum tillage. In response to these needs, he has designed several implements for cotton stalk destruction, drip tubing installation, and tillage operations, some of which have been granted U.S. patents. Mr. Wuertz was Farmer of the Year in 1990, Arizona Farm Bureau Federation, and received the Degree of Doctor of Laws, Honoris Causa, University of Arizona, 1993. Howard Wuertz received a B.S. in agriculture from the University of Arizona in 1951.

Appendix B

Acknowledgments

The preparation of a report such as this one takes input from many people. The committee wishes to extend its sincere appreciation to all the people who shared their time and expertise with us during the study process. This includes the many people who participated in our information-gathering workshop, joined us at meetings in different regions, led us on field trips, helped with our research, and contributed to our study in other ways. In particular, we would like to thank the following people for their contributions:

Gail Achterman, Stoel, Rives, Boley, Jones & Grey, Portland, Oregon
Gene Andreuccetti, Soil Conservation Service, Washington, D.C.
Michael Armstrong, Arkansas Game and Fish Commission, Little Rock, Arkansas
Mark Bennett, Arkansas Soil and Water Conservation Commission, Little Rock, Arkansas
Reed Benson, Oregon Water Watch, Portland, Oregon
Bob Bevis, farmer, Scott, Arizona
Jan Boettcher, Oregon Water Resources Congress, Salem, Oregon
Jerry Butchert, Westlands Water District, Fresno, California
Joe Carmack, Soil Conservation Service, Washington, D.C.
Ken Carver, High Plains Underground Water Conservation District No. 1, Lubbock, Texas
Manzoor E. Chowdhury, Texas A&M University, College Station
Jeb Cofer, Taylor Fulton, Inc., Palmetto, Florida
Dan Daley, Bonneville Power Administration, Portland, Oregon

Al Dedrick, U.S. Water Conservation Laboratory, Phoenix, Arizona
Thomas Donnelly, National Water Resources Association, Arlington, Virginia
Cindy Dyballa, U.S. Department of Environmental Protection, Washington, D.C.
Keith Eggleston, Bureau of Reclamation, Denver, Colorado
Tom Fortner, Soil Conservation Service, Lonoke, Arkansas
Nicky Hargrove, farmer, Stuttgart, Arkansas
Becky Hiers, Umatilla Tribes, Pendleton, Oregon
Peter G. Hubbell, Southwest Florida Water Management District, Brooksville, Florida
Michael Jackson, U.S. Senate Committee on Indian Affairs, Washington, D.C.
Marvin Jensen, consultant, Ft. Collins, Colorado
Stan Jensen, Pioneer Hibred International, Inc., York, Nebraska
Thomas H. Kimmell, The Irrigation Association, Fairfax, Virginia
Frances Korten, The Ford Foundation, New York, New York
Steve Kresovich, USDA Research Leader for Genetic Resources, Griffin, Georgia
Susanne Leckbank, Bureau of Indian Affairs, Phoenix, Arizona
Steve Light, Minnesota Department of Natural Resources, St. Paul, Minnesota
Ronald B. Linsky, National Water Research Institute
Jacke Looney, University of Arkansas, Fayetteville
Curtis Lynn, consultant, Visalia, California
Derrel Martin, University of Nebraska, Lincoln
J. William McDonald, Bureau of Reclamation, Denver, Colorado
Lauren McDonald, White River Valley Association, Newport, Arkansas
Clifton Meador, Arkansas Development Finance Authority, Little Rock, Arkansas
Furman Peebles, Pine City Farms, Rochelle, Georgia
Dean Pennington, Yazoo Mississippi Delta Joint Water Management District, Cleveland, Mississippi
Richard Pinkham, Rocky Mountain Institute, Snowmass, Colorado
Charlie Pintler, farmer, Nampa, Idaho
Pepper Putman, The Irrigation Association, Fairfax, Virginia
Bob Reuter, Westlands Irrigation District, Hermiston, Oregon
Ardell Ruiz, Intertribal Agriculture Council, Sacaton, Arizona
Anil Rupasinghe, Texas A&M University, College Station
Eduardo Segarra, Texas Tech University, Lubbock, Texas
Bill Shepard, rancher, Fallon, Nevada
Earl Smith, Arkansas Soil and Water Conservation Commission, Little Rock, Arkansas
G. Stephen Smith, Sullivan & Associates, Lonoke, Arkansas
Randall Stocker, Imperial Irrigation District, Imperial, California
Gene Sullivan, Sullivan & Associates, Lonoke, Arkansas
Raymond Supalla, University of Nebraska, Lincoln

A. Dan Tarlock, Illinois Institute of Technology, Chicago-Kent College of Law, Chicago
Jan van Schilfgaarde, U.S. Department of Agriculture, Beltsville, Maryland
E.D. "Sonny" Vergara, Peace River/Manasota Regional Water Supply Authority, Bradenton, Florida
John Volkman, Northwest Power Planning Council, Portland, Oregon
Lori H. Walton, White River Regional Irrigation District, Stuttgart, Arkansas
Darrell Watts, University of Nebraska, Lincoln
Paul Westerfield, U.S. Geological Survey, Little Rock, Arkansas
Eric Wilkinson, Northern Colorado Water Conservancy District, Loveland, Colorado
Stuart Woolf, Woolf Enterprises, Huron, California
Randy Young, Arkansas Soil and Water Conservation Commission, Little Rock, Arkansas
David Zilberman, University of California, Berkeley

Appendix C

Glossary

ACRE-FOOT—A traditional measure of water applied, used in the United States. The volume of water required to cover 1 acre of land to a depth of 1 foot. Equal to 1.2 megaliters (ML) or 1,233.5 cubic meters (m^3).

ANADROMOUS—Fish species that ascend rivers from the sea to breed.

APPROPRIATION DOCTRINE—The system of water law dominant in the western United States under which (1) the right to water was acquired by diverting water and applying it to a beneficial use and (2) a right to water acquired earlier in time is superior to a similar right acquired later in time. Also called prior appropriation doctrine. In most states, rights are not now acquired by diverting water and applying it to a beneficial use. Such a system is referred to as the constitutional method of appropriation. Rights are acquired by application, permit, and license, which may not require diversion and application to a beneficial use. Superiority of right is based on earliest in time and has no reference to whether two rights are for a similar use.

AQUIFER—A formation, group of formations, or part of a formation that contains sufficient saturated permeable material to yield significant quantities of water to wells and springs.

BENEFICIAL USE—A use of water resulting in appreciable gain or benefit to the user, consistent with state law, which varies from one state to another.

CENTER PIVOT IRRIGATION—An irrigation system that pumps ground water from a well in the center of a field through a long pipe, elevated on wheels, that pivots around the well and irrigates the field in a large circular pattern.

CONSUMPTIVE USE—Use of water that renders it no longer available because it has been evaporated, transpired by plants, incorporated into products or crops, consumed by people or livestock, or otherwise removed from water supplies.

DEPLETION—Net rate of water use from a stream or ground water aquifer for beneficial and nonbeneficial uses. For irrigation or municipal uses, the depletion is the headgate or wellhead diversion minus return flow to the same stream or ground water aquifer.

DIVERSION—A turning aside or alternation of the natural course of a flow of water, normally considered physically to leave the natural channel. In some states, this may be a consumptive use direct from a stream, such as by livestock watering. In other states, a diversion must consist of such actions as taking water through a canal, pipe, or conduit.

DRIP IRRIGATION—A form of microirrigation.

EVAPOTRANSPIRATION—The sum of evaporation and transpiration from a unit land area. Also see consumptive use.

FLOOD IRRIGATION—A surface irrigation system where water is applied to the entire surface of the soil so it is covered by a sheet of water; called "controlled flooding" when water is impounded or the flow is directed by border dikes, ridges, or ditches.

FURROW IRRIGATION—A surface irrigation system where water is directed down furrows between rows of crops. Common for annual row crops, trees, and vines.

GROUND WATER OVERDRAFT (MINING)—The withdrawal of ground water through wells, resulting in a lowering of the ground water table at a rate faster than the rate at which the ground water table can be recharged.

HIGH-VALUE CROPS (SPECIALTY CROPS)—Crops with a limited number of producers and demand or those with high per acre production costs and value. Examples include most fruit and vegetable crops, ornamentals, greenhouse crops, spices, and low-volume crops such as artichokes.

INPUTS—Items purchased to carry out a farm's operation, such as fertilizers, pesticides, seed, fuel, and animal feeds and drugs.

INSTREAM USE—Any use of water that does not require diversion or withdrawal from the natural watercourse, including in-place uses such as navigation and recreation.

IRRIGATION—The application of water to soil for crop production or for turf, shrubbery, or wildlife food and habitat. Intended to provide water requirements of plants not satisfied by rainfall.

IRRIGATION DISTRICT—In the United States, a cooperative, self-governing public corporation set up as a subdivision of the state, with definite geographic boundaries, organized to obtain and distribute water for irrigation of lands within the district; created under authority of the state legislature with the consent of a designated fraction of the landowners and having taxing power.

IRRIGATION EFFICIENCY—Ratio of irrigation water used in evapotranspiration to the water applied or delivered to a field or farm. This is one of several indices used to compare irrigation systems and to evaluate practices.

IRRIGATION FREQUENCY—Time interval between irrigations.

IRRIGATION RETURN FLOW—The part of applied water that is not consumed by evapotranspiration and that migrates to an aquifer or surface water body. See also return flow.

IRRIGATION WATER REQUIREMENT—The quantity, or depth, of water, in addition to precipitation, required to obtain desired crop yield and to maintain a salt balance in the plant root zone.

IRRIGATION WITHDRAWALS—Withdrawal of water for application on land to assist in the growing of crops and pastures or to maintain recreational lands or landscaping.

LAND LEVELING—Earth moving done on irrigated fields to improve surface slope and smoothness to facilitate water application. Land leveling, which can include laser leveling, can produce uniform slopes in the direction of irrigation-stream advance and may improve conditions for salinity control by improving uniformity of surface irrigation.

MEGALITER—A measure of volume: 1 ML equals 0.8107 acre-foot; 1 acre-foot equals 1.23 ML and is the volume of water required to cover 1 acre of land to a depth of 1 foot.

MICROIRRIGATION—Irrigation methods such as drip/trickle, subsurface, bubbler, and spray irrigation. In microirrigation systems, water usually is delivered to the soil near the plants through a network of tubing with closely spaced, low-flow rate emitters. Water typically infiltrates where applied, and the soil volume wetted is therefore controlled by the number of application points and the lateral movement of water in the soil. Systems of emitters are easily automated, making frequent, light water applications possible without high labor costs. A high level of management is needed.

NET DEPLETION—Total water consumed by irrigation, or other use in an area, equal to water withdrawn minus return flow.

PHREATOPHYTE—A deep-rooted plant that obtains its water from the water table or the layer of soil just above it.

PRIOR APPROPRIATION—A concept in water law under which a right is determined by such a procedure as having the earliest priority date.

PUBLIC INTEREST—An interest or benefit accruing to society generally, rather than to any individuals or groups of individuals in the society.

PUBLIC TRUST DOCTRINE—A poorly defined judicial doctrine under which the state holds its navigable waters and underlying beds in trust for the public and is required or authorized to protect the public interest in such waters. All water rights issued by the state are subject to the overriding interest of the public and the exercise of the public trust by state administrative agencies.

REACH—A specified length of a stream or channel.

RETURN FLOW—The amount of water that reaches a ground or surface water source after release from the point of use and thus becomes available for further use. See also irrigation return flow.

RIPARIAN—Relating to or living or located on the bank of a natural watercourse, usually rivers but sometimes lakes or tidewaters.

RIPARIAN RIGHTS—A concept of water law under which authorization to use water in a stream is based on ownership of the land adjacent to the stream and normally not lost if not used.

RUNOFF—That part of the precipitation that moves from the land to surface water bodies.

SALINIZATION—To become impregnated with salt; concentration of dissolved salts in water or soil water. An environmental impact of irrigation that can be managed but not eliminated.

SPRINKLER IRRIGATION—Sprinkler irrigation systems can be either set or mobile. In set systems, the sprinkler heads are stationary while applying water. Mobile systems move continuously, either in straight lines or circular patterns while irrigating; these generally cost more than set systems but require less labor.

TRANSFER (CONVEYANCE OF WATER RIGHT)—A passing or conveyance of title to a water right; a permanent assignment as opposed to a temporary lease or disposal of water.

WATERSHED—A geographic region (area of land) within which precipitation drains into a particular river, drainage system, or body of water that has one specific delivery point.

WATER USE EFFICIENCY—Marketable crop production per unit of water consumed in evapotranspiration.

WATER WITHDRAWAL—Water removed from ground or surface water sources for use.

Index

C

J

K

L

M

Q

R

S

X

Y